# THE EVERYDAY GUIDE TO BRITISH BIRDS

**Charlie Elder**

BLOOMSBURY WILDLIFE
LONDON • OXFORD • NEW YORK • NEW DELHI • SYDNEY

**For my brother and sisters Jay, Charlotte and Tamsin**

BLOOMSBURY WILDLIFE
Bloomsbury Publishing Plc
50 Bedford Square, London, WC1B 3DP, UK

BLOOMSBURY, BLOOMSBURY WILDLIFE and the Diana logo are trademarks of Bloomsbury Publishing Plc

First published in the United Kingdom 2018



A catalogue record for this book is available from the British Library.
Library of Congress Cataloguing-in-Publication data has been applied for.

ISBN: PB: 978-1-4729-8762-4; ePDF: 978-1-4729-4115-2; ePub: 978-1-4729-4116-9

2 4 6 8 10 9 7 5 3 1

Design by Rod Teasdale
Printed and bound in China by C&C Offset Printing Co., Ltd.

To find out more about our authors and books visit www.bloomsbury.com and sign up for our newsletters

Published under licence from RSPB Sales Limited to raise awareness of the RSPB (charity registration in England and Wales no 207076 and Scotland no SC037654).

For all licensed products sold by Bloomsbury Publishing Limited, Bloomsbury Publishing Limited will donate a minimum of 2% from all sales to RSPB Sales Ltd, which gives all its distributable profits through Gift Aid to the RSPB.

# Contents

# Introduction

Being able to put a name to the birds you come across on a walk, or spot through the kitchen window can be hugely satisfying, adding immeasurably to the pleasure of being outdoors and the enjoyment of attracting feathered friends to your garden. *The Everyday Guide to British Birds* is written for nature enthusiasts and birdwatching beginners who want to get to grips with our most abundant species and learn more about what makes each of them special.

While the UK plays host to around 280 regularly-occurring bird species, only serious devotees could hope to see them all, given a significant number are rare or hard to find. *The Everyday Guide to British Birds* keeps identification simple by focusing on probabilities rather than innumerable possibilities. It takes as its starting point, 80 common and widespread birds that we're most likely to encounter in Great Britain and also flags up similar-looking species, some scarcer, that could cause confusion.

All the main birds featured are found across much of Britain in their favoured habitats. The first half of this guide covers birds of woods, fields, gardens and wilder countryside, while the second half focuses on species generally associated with water, including lakes, rivers, estuaries and the sea. In addition, birds are grouped together in sections based on shared characteristics, such as 'birds of prey' or 'ducks, geese and swans', rather than in

strict taxonomical order. Not all birds can be neatly pigeonholed, however these broad groupings aim to provide readers with a common sense starting point when they're trying to identify a sighting.

Many birds call the UK home all year round, like Wrens and Blue Tits. Others come here during the winter, to avoid freezing conditions further north, including various geese and waders; while species such as Swallows travel from southerly climes to breed in the spring and summer, taking advantage of our long days and plentiful insect life. The picture changes with the passing seasons. This book indicates when, as well as where, you are likely to see our most familiar birds, along with key plumage details that may differ depending on the species' age, sex or the time of year.

Of course, there is more to birdwatching than simply sticking name labels on everything you see and *The Everyday Guide to British Birds* includes interesting facts about the featured species and descriptions of the qualities that make each of them unique.

Finally, a word of reassurance for those just beginning birdwatching. Stepping out into the countryside shouldn't feel like setting foot in an exam room. It really doesn't matter if you can't recognise a species or you make mistakes with identification. When it comes to our diverse birdlife, no one has seen it all, or knows it all – there is simply too much to experience and learn. Likewise, no one knows nothing at all about birds. We are all bumping along somewhere in the middle. Enjoy.

# Woodland, parks and gardens

# Wren

## Key identification features

■ The Wren is a very small, rotund, brown bird most often seen foraging at low level in a rather mouse-like fashion.

■ It is darker on top and frequently holds its short tail cocked vertically. With a good view, subtle barring on the sides and a pale line above the eye can be made out. Wrens generally stick close to cover, flying short distances on whirring rounded wings.

■ The affectionate nickname 'Jenny Wren' is applied to both females and males because it is virtually impossible to tell the sexes apart.

## Where could you find one?

Pretty much everywhere. The Wren is one of our most widespread birds, at home anywhere with plenty of vegetation. There is no risk of getting a crick in your neck searching treetops for a sighting: Wrens exist comfortably below eye level, at home among thickets, at the base of hedges, and in the nooks and crannies of stone walls and tree roots. They don't tend to visit bird tables, though might sneak out for tiny titbits on the ground.

## Is it easy to identify?

The Wren is not necessarily a timid bird, but a combination of smallness and fairly drab plumage means this common species can be easy to overlook. Its restless nature may give away its presence in hedgerows and along overgrown walls, and when its feathers are fluffed up, this neckless ball of a bird, with thin bill and stubby tail, has something of a tiny teapot shape. For a species of its size, the Wren delivers an extraordinarily loud and attention-grabbing song. The brief and strident torrent of notes, incorporating a tell-tale trill, can burst forth from deep within a bush and stop you in your tracks as you try to figure out what could be making such a sound – only to marvel as the perky little performer hops into view.

## How likely am I to see one?

Small yet vocal

Though not as conspicuous as the House Sparrow, Robin or Blackbird, the Wren is Britain's most common bird, with some 7.7 million pairs widely distributed across the country. As Wrens generally keep a low profile, one is as likely to be alerted to their whereabouts by their calls and song as their movement.

## What makes it special?

Small in size, but with a larger-than-life character, the Wren has won a place in our hearts, coming fourth in a 2015 poll to choose the UK's national bird. The Wren is undeniably cute and astonishingly vocal: the volume of both its high-pitched song and sharp alarm calls seemingly defy the physical limitations of a bird weighing much the same as a pound coin. Endearing and enterprising, this familiar feathered friend has captured imaginations down the centuries, attracting its fair share of folklore. One well-known fable describes how the diminutive Wren was crowned king of the birds in a competition between species to fly the highest, by hiding within the plumage of an Eagle and taking to the wing once the soaring raptor could climb no further. During cold winters its population suffers huge losses, but resolutely bounces back as hard-working pairs are able to raise up to a dozen young a year when times are good.

## Added interest

- Male Wrens can have several mates and build half a dozen or so ball-shaped mossy nests tucked away in banks, walls or foliage, from which a female takes her pick, finishing it off with a feather lining.
- In hard winters, Wrens will roost together, huddling up side-by-side in old nests or cavities to share warmth. A staggering 63 have been recorded sharing a single nest box!
- Despite its size, the Wren is not our smallest bird – the Goldcrest is the littlest of the lot, weighing a mere 6.5g and laying eggs just 14mm long.

## Not quite a Wren?

If your tiny Wren has a dark-bordered yellow stripe on its rather plain head, then this is almost certainly a Goldcrest. This widespread species, numbering over 500,000 pairs in Britain, favours conifers but can be spotted pretty much wherever there are bushes and trees. Goldcrests have dull green plumage, a short white bar visible on the folded wing and a yellow or orange crest. If, on the other hand, your Wren looks a little on the large side and has a greyish head then it could be a Dunnock (*see Dunnock*). Finally, if it appears distinctly pale underneath, with a longer down-curved bill and is happily working its way vertically up the trunk of a tree then it could be a Treecreeper (*see Nuthatch*).

Goldcrest

# Blue Tit

## Key identification features

- The Blue Tit is a familiar garden visitor: an agile little bird with a blue cap and yellow belly.
- Smaller than a Sparrow, it has a white face surrounded by a dark blue collar and a black line runs through the eye. Its wings, tail and top of the head are bright blue, while its back has a dull greenish tinge.

## Where could you find one?

Anywhere where there are trees and bushes, across almost all of Britain throughout the year. Deciduous woodlands are the habitat of choice, providing plenty of insect food during the warmer months and holes in which to nest, but the adaptable Blue Tit is also abundant in parks and gardens and can be found in conifer forests. It readily comes to bird tables, particularly in the autumn and winter when insect food is hard to find, and uses nest boxes with a small hole entrance.

## Is it easy to identify?

The Blue Tit is our only small bright blue bird, except the Kingfisher (which you won't find hanging on your peanut feeder), making identification fairly straightforward. The overall impression is small, light blue, white and yellow – and if it stays still for a moment then the elements become more obvious: light blue beret, white cheeks and pale lemon belly. Where it becomes harder to identify is out and about in woodland feeding high among the outermost branches of trees. Its size, rotund neckless shape and restless nature, points to it being some kind of tit but the azure cap sets it apart from close relatives. It is a vocal bird and its song is a quick high-pitched ditty, typically starting with two or three longer notes followed by a trill, similar in Morse Code rhythm to a couple of 'dashes' then a hurried succession of 'dots'.

## How likely am I to see one?

Regular feeder visitors

Hang a peanut feeder or fat balls outside your window and it shouldn't take long! Blue Tits are common garden visitors that are quick to take advantage of free food hand-outs and nest boxes. There are an estimated 3.5 million pairs living in all kinds of wooded habitat across almost all of Britain (apart from some of the Scottish islands) and they are one of the most frequently spotted species in the RSPB's annual Big Garden Birdwatch.

## What makes it special?

The Blue Tit is understandably a national favourite. Cute, colourful and characterful, the species rewards our interest in providing for nature by willingly using nest boxes and is an enthusiastic visitor to bird feeders – even those stuck to kitchen windows. It brightens up bird tables with acrobatic feeding antics, while family dramas streamed live on TV from 'nest-cams' have captivated audiences. Not only is the dedication of parent birds endearing, it is also impressive – these chipper little bundles of brightness are hard workers. A pair typically lays one clutch averaging eight to ten eggs, timed to coincide with a spring glut of insect life, in particular moth caterpillars. Satisfying the demands of a large and hungry brood means continuous trips for parents between treetops and the nest, delivering caterpillars to gaping chicks at the rate of a visit every ninety seconds.

## Added interest

- Blue Tits put all their eggs in one basket, so to speak, during the breeding season. Instead of having several broods, they lay a single clutch which can number up to 16 eggs – a staggering total for such a small bird.
- Feeding a large number of young is exhausting work for parents, who can jointly bring up to 1,000 caterpillars to a nest every day.
- Blue Tits can see ultraviolet wavelengths of light. While the sexes look similar to us, the male's blue crown shines brightly in UV light and helps determine his attractiveness to potential mates.

## Not quite a Blue Tit?

If it is summer and your Blue Tit is looking a little washed out, then it is probably a juvenile – they have a yellowish face and the belly and cap are duller than in adults. If it has a black cap, then it could be one of the other tits. The larger Great Tit has a black cap and collar and a black stripe running down the centre of its golden front (*see Great Tit*), while the other tits lack the yellow bellies of the Blue and Great Tit (*see Coal Tit*). Another similar-sized bird that can be seen feeding among the outer branches of trees is the Goldcrest, which is greenish-brown and has a blaze of colour on the head (*see Wren*).

Great Tit

# Great Tit

Female Male

## Key identification features

- The Great Tit is about the size of a house sparrow and the largest member of the tit family.
- It has a black head, throat and collar with a prominent white cheek patch, a greenish back and blue-grey wings. The front is bright yellow and has a black stripe running down the centre of the breast and belly, which is thicker in the slightly smarter-looking males.
- The Great Tit is a vocal bird, with a range of calls and songs.

## Where could you find one?

They are primarily birds of deciduous woods (especially those with oak trees) but can also be found in copses, conifer plantations, hedgerows, parks and gardens. Widespread across Britain throughout the year, Great Tits seldom travel far, holding a territory during the breeding season and joining other birds roaming through woods in search of food during the colder months. They visit garden bird feeders, especially in the winter, and will readily use hole-fronted nest boxes.

## Is it easy to identify?

The bold segments of colour make the Great Tit a fairly straightforward bird to identify and it can be easy to view as it feeds on and around bird tables. The smaller Blue Tit also has a yellow front, but not the black head markings or conspicuous inky brushstroke running down the front – which gives the Great Tit the appearance of having donned a dapper yellow waistcoat. In addition, the Great Tit draws attention to itself with a remarkable range of peeping or churring calls and song variations and is among the earliest in the spring to start singing. Its most distinctive song is a high-pitched strident see-sawing ditty which sounds like it is saying *teacher, teacher, teacher*.

## How likely am I to see one?

Attracted to gardens

Great Tits are common birds across Britain throughout the year, with an estimated 2.5 million pairs living in our woodlands and gardens. Hole-fronted nest boxes and bird feeders are almost certain to attract them to your garden. In keeping with other small birds, their average life expectancy is short, however there are records of individuals surviving over thirteen years!

## What makes it special?

The Great Tit is rightly a popular garden bird, with its handsome plumage, confident character and willingness to use feeders and nest boxes. It would be easy to take its looks for granted, given it is such a common species, but in the feather fashion stakes, this backyard regular could arguably hold its own in the tropics among the showiest of small species. A glossy black head with eye-catching white cheeks set it apart, while the smartly contrasting black and yellow on the front is undeniably fetching and plays a part in impressing potential mates. The brightness of an individual's yellow plumage is a sign of its good health and adeptness at finding invertebrate food, as the pigment is derived from eating caterpillars. Great Tits are excellent pest controllers: each pair collects hundreds of caterpillars a day to feed to hungry broods of, on average,

seven to nine chicks. They are resourceful and intelligent. At a time when doorstep deliveries of full-fat milk were commonplace, they joined Blue Tits in working out how to peck open foil bottle tops to get at the digestible cream on top. Researchers have found that the ability to solve such brain-teasing challenges quickly spreads through a population as these colourful and canny birds observe and learn from one another.

## Added interest

- The Great Tit's original family name 'Titmouse' describes the busy little appearance of the various tit species and derives from early English words for 'small'.
- The striking black stripe down the front of Great Tits is more well-defined in males, running in a broad band between the legs and is believed to be a factor in attracting females.
- Great Tits living in towns and cities have been found to communicate at a higher pitch than those in rural areas, in order to overcome background traffic noise.

## Not quite a Great Tit?

If your Great Tit looks on the small side and has blue on the head rather than black, then it's a Blue Tit (*see Blue Tit*). If, however, you spot a small bird with a black cap and white cheeks on the bird feeder, or roving among the branches of a tree, which lacks the bright yellow underparts of the Great Tit, then look carefully at the back of the head – a white stripe at the rear of the black cap is the giveaway sign of a Coal Tit (*see Coal Tit*). Finally, if your Great Tit is looking a little dull and has yellowish cheeks then it could be a juvenile which has yet to assume the vivid colours of an adult.

Coal Tit

# Coal Tit

## Key identification features

- The Coal Tit is the tiniest of the tit family; an active and acrobatic little bird with a small body, short tail and thin dark beak.
- It has a proportionately large black head with white cheek patches and a diagnostic white stripe runs down the rear of the crown. The underside has a buff tinge and the back is grey. Two short white lines are noticeable dotted along the side of the closed wing.

## Where could you find one?

Evergreen forests are the habitat of choice for this widespread and common species, which has benefitted from the planting of commercial conifer stands. However, they can also be found in mixed woodland, parks and gardens, where they will visit bird feeders and sometimes use hole-fronted nest boxes if larger Great Tits and Blue Tits don't occupy them first. They seldom travel far and are found pretty much across the UK year-round in suitable wooded areas.

## Is it easy to identify?

The Coal Tit's most obvious feature is best seen from the rear: a badger-like white stripe down the back of the black head. Even when they are feeding high in the canopy, or darting at speed to and from a bird table, this short chalk line stands out. It also helps distinguish the Coal Tit from the larger Great Tit – which has a similar black head and white cheek patches, though a yellow front with a black stripe running down the centre. Coal Tits are vocal and their high-pitched calls draw attention to their presence as they forage among tree branches for invertebrates during the summer and seeds in winter. They also have a simple song that sounds a bit like a thin, fast version of the Great Tit's rhythmic *teacher, teacher, teacher* ditty.

## How likely am I to see one?

A conifer specialist

This is a common little bird, particularly so where there are conifer trees. There are an estimated 680,000 pairs in Britain, with the highest concentrations in Scotland and Wales, as well as the south and west of England. Bird tables, particularly in rural areas or near woods, will attract Coal Tits, which can become regular visitors in winter.

## What makes it special?

The Coal Tit offers something a little different from the more common Great Tit and Blue Tit. However, being so small, it typically loses out to its larger cousins at the bird table and so tends to make more fleeting grab-and-dash visits for peanuts, sunflower seeds and other snacks - opting for take-out rather than dining in. Coal Tits seek safety in numbers outside the breeding season by joining other small bird species in feeding flocks that roam through woods and hedgerows. Being so light, they can hang upside down on the thinnest of twigs to find food, even the underside of snow-laden branches, and will hover briefly to pick off insects. They would be easy to miss, but their high squeaky calls give them away, and in conifer plantations, Coal Tits can be a welcome sign of life among the otherwise empty-seeming corridors of trees.

## Added interest

- The Coal Tit tends to nest in tree holes low down to the ground and will even use mouse burrows or crevices among roots in which to raise young.
- Getting at insects and seeds tucked away within pine cones, clusters of pine needles or bark crevices isn't easy, but this conifer-specialist has a slender pointed bill perfectly adapted for the job.
- In times of plenty, Coal Tits stash food under moss, bark or lichen to eat later, which is why they frequently make repeated short-stop visits to garden bird feeders.

## Not quite a Coal Tit?

If your Coal Tit lacks the white stripe down the rear of its black head, then you could be looking at an exciting find - the Marsh Tit. This woodland-dweller, numbering 41,000 pairs across England, Wales and southern Scotland, is quite similar to the Coal Tit, though with no little white dotted bars on the wings. Confusingly, the Marsh Tit has an even scarcer lookalike: the Willow Tit. This declining species, of which there are only 3,400 pairs in Britain, is virtually identical, though its cheeks are a cleaner white throughout and its neck is thicker. The most reliable distinction is that the Marsh Tit makes a sneezing pit-choo sound and it has a small white mark near the base of the beak's upper mandible.

Marsh Tit

# Long-tailed Tit

## Key identification features

- The Long-tailed Tit is a delightful little ball of a bird with an elongated narrow tail which makes up more than half its length.

- It is mostly black on the back, wings and tail, and light underneath. Its face is pale, with dark eyebrows that emphasise a white crown and it has a pinkish flush on the underside and some darker pinkish-brown feathering on the shoulders. It also has a very short, stubby beak.

## Where could you find one?

This is a widespread and fairly common British resident which inhabits deciduous woodlands, hedgerows, parkland and scrubby areas where it can find a plentiful supply of insect food, places to nest and thick cover in which to hide and roost. It is also becoming increasingly familiar in urban and suburban areas and at bird feeders.

## Is it easy to identify?

The lollipop shape is the giveaway: a rotund little body and long stick tail. These features are obvious even in flight as roving flocks flit on short rounded wings between trees – the tail noticeably longer than the body. From a distance the Long-tailed Tit looks quite black and white, but a closer view shows a wash of pale candyfloss pink on the underside and a patch of darker pink on the shoulders. They are a very gregarious species, particularly outside the spring breeding season and are invariably spotted in groups of between six and 20 birds. Members of these loose flocks, which can forage quite close to human observers, constantly keep in touch with one another as they go, making high-pitched see-see-see calls and purring trills.

## How likely am I to see one?

Active feeders

Long-tailed Tits are unpredictable birds that can be all around you one minute and gone the next. However, they seldom travel far, remaining within woodland territories and so, in suitable habitat, one is likely to come across them regularly. They are not uncommon, given there are an estimated 330,000 pairs across Britain, but always feel like a treat.

## What makes it special?

There are few birds quite as endearing as the Long-tailed Tit. Not only is this golf ball-sized bird attractively patterned and flushed with pink, but its idiosyncratic tail, bright eyes, acrobatic feeding style and sociable nature adds an undeniable charm. In the depths of winter, Long-tailed Tits will huddle together side-by-side to share body warmth – a cluster of fluffy puffballs with straight tails sticking out. Cute as it may sound, this is the time of year that takes its greatest toll on numbers of this small insect-eating species and bird feeders offering suet and small seeds can be a lifesaver. During the spring breeding season, pairs laboriously build oval nests of moss, cobwebs and lichen in the fork of a tree or within a low thorny bush. The intricate sack-like construction has a hole in the side and is elastic enough to accommodate a growing brood of, on average, six to eight young.

## Added interest

- Long-tailed Tits can line their nests with over 1,500 downy feathers – many plundered from the corpses of dead birds.
- Despite being well camouflaged, the nests are frequently predated and adults, in particular males, who fail to raise young will instead help feed the broods of neighbouring pairs.
- Flocks commonly seen in autumn and winter are made up of extended family groups, consisting of parents, their offspring and adult siblings who may have helped with nestling feeding duties.

## Not quite a Long-tailed Tit?

If it is summer and the Long-tailed Tit you have spotted is the standard shape and size but has a dark greyish face and lacks any pink feathering, then it could be a juvenile. By the autumn young Long-tailed Tits will have moulted into full adult plumage. And if your Long-tailed Tit looks a little larger than expected, is very black and white and is walking on the ground bobbing its lengthy tail as it goes, then it could be a Pied Wagtail (*see Pied Wagtail*).

Left: Pied Wagtail.
Above: Juvenile Long-tailed Tit

# House Sparrow

Female

## Key identification features

- The House Sparrow is a familiar small, brown, chirpy bird. The males have a streaky chestnut brown back, a short light wing bar, are pale grey underneath and have reddish brown on the sides of the head, a grey cap and black bib.
- The females are mostly light grey-brown, streaky on the back and have a pale stripe running back from above the eye.

## Where could you find one?

Just outside your window, cheeping on a wall, tucking into breadcrumbs on the bird table, nesting under roof tiles, hiding in thick garden bushes, grubbing around for seeds along a field edge, bickering in a hedgerow, scrounging for hand-outs by a park pond, chasing insects in an overgrown verge, loafing around on an industrial site, dust-bathing in a farmyard… pretty much wherever we live, they live, across the UK, all year round.

## Is it easy to identify?

Flocks of these backyard companions are such an everyday sight that we hardly need a second glance to know what they are. House Sparrows are never far from human habitation: gregarious groups flying on whirring wings, cheeping from rooftops and feeding on the ground with short-legged hops. In urban settings House Sparrows can look quite scruffy, as if camouflaged to blend in with dusty paving slabs, grubby brickwork and soot, though a closer look reveals subtle patterning on the back and head and an impressive thick seed-cracking beak. During the spring breeding season, the male's bill is darker and his black 'beard' is much larger – in good light he is quite a handsome chap. In rural settings, a closer look at the head markings is worthwhile as you may have stumbled across the House Sparrow's far scarcer country cousin, the Tree Sparrow (*see right*).

## What makes it special?

The House Sparrow would be easy to take for granted, but has suffered such a catastrophic population decline over recent decades that it has been included on the UK's Red List of species of greatest conservation concern (*see page 186*). The reasons are unclear and the finger of blame has been pointed at a variety of factors, including changing agricultural practices, rises in city pollution and a reduction in suitable breeding sites. House Sparrows nest in cavities, such as those under roofs, are predominantly seed eaters and raise their young on invertebrates. Reduce the availability of some, or all, of these key requirements and numbers are likely to tumble. While still one of Britain's most abundant species, the House Sparrow population is a far cry from what it once was. The centre of London, for example, is now virtually devoid of the formerly abundant 'cockney sparrer'. This resourceful species has lived alongside us for millennia, sharing the crumbs from our table and making its home among our homes. Despite unassuming looks and monotonous calls, these industrious birds brighten our lives with their company and are worthy of appreciation.



### How likely am I to see one?

Numbers of House Sparrows may have plummeted by more than 65 per cent since the 1970s, but they are still widespread and common, with more than five million pairs in Britain. They are the species we are most likely to see from our kitchen window and have remained at top spot in the annual RSPB Big Garden Birdwatch. Despite this, they are not the UK's most abundant bird: the Wren, Chaffinch and Robin are more common.

**The sociable House Sparrow has suffered declines**

## Added interest

- While House Sparrows look like Finches, they are more closely related to Weaver Birds - though their rather untidy domed nests of dry grasses and feathers are far less sophisticated.
- Among the more bizarre places colonies have become established was a south Yorkshire colliery where several lived 640 metres below ground and were fed by miners.
- While pairs stay together for life, infidelity among colonies is rife and DNA studies show a fifth of nests contain one or more chicks unrelated to the main male - who actually provides less food if he believes he is being cheated.

### Not quite a House Sparrow?

If your House Sparrow has a rich chestnut-brown cap and a black patch on its white cheeks, then you're looking at its scarcer, rural relative: the Tree Sparrow. Numbering 180,000 pairs, these birds are mainly found in arable areas of lowland eastern England. Their population declined alarmingly due to agricultural changes, but appears to have turned a corner.

Others that are confused with the House Sparrow include the Dunnock, a common, streaky grey-brown garden species which has a much thinner bill (*see Dunnock*); the female Chaffinch, a more uniform brown with prominent white markings on the wings (*see Chaffinch*); and the female Reed Bunting, which has a hint of a white moustache and white sides to the tail (*see Reed Bunting*).

**Tree Sparrow**

# Dunnock

## Key identification features

■ The Dunnock is a rather plain looking, unobtrusive, Sparrow-sized species with a streaky brown back and a dusting of grey on the head, neck and throat. It generally shuffles around at ground level close to or within cover, though occasionally sings out in the open on perches, such as at the top of a bush.

■ While also known as the 'Hedge Sparrow', the Dunnock lacks the thick seed-crunching beak of Sparrows and Finches and, instead, has a more delicate bill used for picking up insects and titbits.

## Where could you find one?

Wherever there are bushes and trees, hedges and tangles of brambles, bracken or scrub there should be Dunnocks about. This widespread and common species inhabits a range of habitats, from woodland to farmland, just so long as there is enough cover to keep out of view when required as they search for food. They are regulars in parks and gardens, where they will pick up fallen crumbs and seeds from beneath the bird table.

## Is it easy to identify?

The Dunnock can go largely unnoticed, given its plumage lacks eye-catching features and it generally avoids showy behaviour. But the fact that this bird doesn't stand out in a crowd may actually help identify it. Unlike some conspicuous small species with bright feathering, social natures and perky demeanours, the nervy Dunnock tends to move with low hops, creeping about on the ground, skirting the edge of foliage or flitting hurriedly into it when disturbed, without drawing too much attention to itself. Its streaky brown body resembles that of a Sparrow, but the thin bill and smoky grey head set it apart.

## How likely am I to see one?

Warbling ditties

The Dunnock may not be a bird that readily makes itself known, but it is an abundant and widely distributed resident species, numbering some 2.3 million pairs in Britain. The best views can often be had in gardens as they feed along the edges of borders or around bird feeders. They are most likely to put their slightly agoraphobic tendencies to one side when singing, perching out in the open as they deliver their hurried warbling ditties.

## What makes it special?

The Dunnock could rightly be described as an 'LBJ' – the light-hearted birdwatching term for any nondescript 'little brown job'. In fact, its name dates back centuries and literally translates 'little brown bird'. Once you have cracked identifying a Dunnock, it offers the lifelong pleasure of being an LBJ you can actually put a name to. Added to which, it has a certain understated charm all of its own and brightens winter days with short outbursts of cheery warbling. Unusually, lookalike males and females both defend territories, sing and, come the spring, they engage in a complicated assortment of pairings ranging from simple monogamy to multiple mates. Their nests are a favoured target of Cuckoos, which lay their eggs among Dunnock clutches, leaving the hapless hosts to raise their outsized offspring. While Cuckoos cleverly match the shell colour of various bird species in order that their eggs go unnoticed, they fall short when it comes to replicating the bright blue eggs of Dunnocks. In spite of this, Dunnock parents have yet to cotton on to the dissimilarity and happily incubate the imposters' speckled-brown eggs, seemingly none the wiser.

Raising a young Cuckoo



## Added interest

- During the spring and summer breeding season Dunnocks mate frequently, but in the blink of an eye – couplings take just a fraction of a second.
- Any Dunnock males that father offspring will share in the task of raising the young, even when females have multiple mates.
- The Dunnock is a one of a kind species in Britain: our only member of the accentor bird family, which is typically associated with mountainous regions of northern Europe and Asia.

## Not quite a Dunnock?

Plenty of small brown birds could be confused with a Dunnock, from female House Sparrows, sharing the streaky brown back (*see House Sparrow*) and Wrens, with short cocked tails (*see Wren*), to juvenile Robins, speckled brown and lacking a red breast (*see Robin*). If your Dunnock's bill is thick, it might be a female Reed Bunting (*see Reed Bunting*). If the bill is thin, but the plumage differs, then it could be a Warbler (*see Chiffchaff*). The difference in appearance between these birds can be subtle and identifying 'little brown jobs' may boil down to how they behave, where and when you spot them or the sounds they make, as much as how they look.

Female Reed Bunting

# Robin

## Key identification features

- The Robin is a familiar small bird whose tameness, smart red breast and sweet song have made it a national favourite. It has a brown back, pale belly and an orangey-red front, fringed with grey, that covers the face and throat, extending in two lobes over the chest like an upside down love heart.

- The Robin has appealing large dark eyes, a short little bill and thin legs. When fluffed up it can appear very rotund. Both males and females look alike, but juveniles are speckled brown and the first signs of red breasts only begin to appear after a couple of months.

## Where could you find one?

Robins are abundant and widespread all year round in habitats such as woods, hedgerows, parks and gardens that have plenty of undergrowth where they can nest low to the ground. Although they will feed at bird tables and eat berries in the autumn, they depend on plenty of insect life throughout the spring and summer to raise two or three broods, catching invertebrates spotted from a perch or by hopping along the ground.

## Is it easy to identify?

Juvenile

A photogenic friend of the gardener and cheery emblem of Christmas, the Robin is one of our most familiar birds. Its breast colouring is perhaps more orange than red, however the word orange did not enter our language as a colour description until the mid-1500s following the earlier introduction of the fruit (before which the colour was described in Old English as 'yellow-red') and the historic term 'redbreast' has stuck. While adults are instantly recognisable front on, with both sexes alike, the scruffy, mottled-brown juveniles initially lack a red breast. Robins can also be identified by their warbling song, which is used to signpost territory. Good eyesight means they can be up and singing before sunrise and even after dark by the light of street lamps, when they may be mistaken for the summer-visiting nightingale. Significantly, they are one of only a few birds that sing in the winter.

## How likely am I to see one?

Moulting to replace old feathers

Not only is the species abundant, widespread and present throughout the year (with an estimated six million pairs in Britain), resident Robins seldom venture more than a few kilometres during their entire lives. However, a few do cross over to the Continent in winter, while Robins from Scandinavia can turn up on our east coast. Robins can be seen and heard in every season, though in late summer, when they moult their feathers, they tend to keep a low profile.

## What makes it special?

In a 2015 poll to choose the UK's national bird, the Robin comfortably came top, attracting more than a third of the votes cast. A part of our everyday lives country-wide, the Robin was protected from persecution down the centuries by folk tales of ill fortune afflicting those who harmed one – which may explain why they can be so tame in Britain. Their perky character, delightful song and endearing looks have won our hand of friendship – especially one filled with seed mix and mealworms. However, the Robin is far from good-natured towards rivals when it comes to defending the precious territory upon which it depends for its survival and breeding success. If angry calls and puffing up the red chest fail to deter an intruder it will resort to full-on assault, with one in ten bouts ending in a fatality. When the red mist descends, it will even attack anything of that colour – including stuffed Robin toys and bunches of red feathers.

## Added interest

- Unusually for songbirds, female Robins sing – in particular during the winter when they hold their own territory.
- Robins have evolved to shadow the ground-grubbing activities of animals such as wild boar and deer, watching for invertebrates turned up in the soil, which is why they are famously attentive around gardeners.
- When it comes to choosing somewhere secretive and sheltered to nest, Robins have opted for some bizarre locations, ranging from discarded kettles and coat pockets, to skulls, plant pots and beneath car bonnets.

## Not quite a Robin?

The Bullfinch, Chaffinch and Stonechat all have rosy or rufous chests, however the Robin has no markings on the head, rump or wings which can help identify these other species (*see Bullfinch, Chaffinch and Stonechat*). The Redstart is also an exciting possibility. This handsome migrant, numbering 100,000 pairs in Britain, is only present during the spring and summer breeding season and males have a grey back, black face and orange-red on the front and tail. Finally, Robins singing from cover as daylight fades may be mistaken for far scarcer Nightingales. These brown, similar-sized summer visitors are found mainly in the south-east of England and their astonishing virtuoso performances include a rich repertoire of complex phrases which are worth getting to know.

Male Redstart

# Chaffinch

Female

## Key identification features

■ The Chaffinch is a colourful, confident and common finch, Sparrow-sized but slimmer, neater and longer-tailed.

■ Males have a mixed plumage palette: an olive-green rump, chestnut-brown back, slate-blue head on the top and back, a pale salmon-pink belly and rusty-reddish brown cheeks, throat and breast.

■ Females are plain by comparison – greyish-olive brown and slightly darker on the back. Both of them have a white patch on the shoulder and a white bar across the wing and also white outer tail feathers.

## Where could you find one?

This abundant and adaptable species can be found in almost any area where there are trees and bushes, from deciduous and coniferous woodlands to farmland hedgerows, scrubland, parks and gardens. They are distributed across the UK all year round and are bird table regulars, typically feeding on the ground for fallen seed.

## Is it easy to identify?

Despite the varied colours of the male, mixing reddish brown, pink, green and blue-grey, the Chaffinch is anything but garish. Feeding along a woodland edge in dappled light the plumage blends in subtly among fallen leaves and bare soil. Even more so the female, whose dull camouflage tones enable her to sit unobtrusively on a nest while incubating eggs. However, both share a standout feature that catches the eye when they move: a bold brushstroke of white on the 'shoulder' of the wing which is particularly visible in flight. It can be seen on the folded wing as well and is accompanied by a white bar running across the flight feathers just beneath it. They are not shy birds and the males are a treat to observe close-up. They are also very vocal, with a range of calls and a short cheerful song.

## How likely am I to see one?

Energetic singer

The Chaffinch is Britain's third most common breeding bird after the Wren and Robin, with nearly six million pairs present in the breeding season. This resident population is boosted in the autumn as Chaffinches from northern Europe - in particular Scandinavia - arrive in Britain, fleeing advancing cold weather and remaining in the country during the winter.

## What makes it special?

If you have yet to get to grips with mastering bird songs, the Chaffinch's ditty offers a satisfying introduction. One step up from *cuckoo* or *quack*, it is complicated enough to challenge and reward the learner, but simple enough to recognise. The springtime refrain is repeated again and again, loudly and clearly, typically out in the open from the top of a bush where you can match the song and singer. It comprises a two-second rattling series of notes which accelerate into a final flourish and has been described as having the rhythm of a cricketer running in and bowling a ball, the pace quickening from a few steady strides into a rapid tempo before the final vigorous delivery. It is well worth listening to a birdsong CD or online in order to recognise it. Only male Chaffinches perform the song, to stake out their territory and attract a mate and, interestingly, it is not only birdwatchers who are keen to learn how it goes. Young male Chaffinches listen acutely to the recurring renditions. They are born with the ability to sing the song, but need to study exactly how it is structured and practise before they are ready to breed the following spring.

## Added interest

- The Chaffinch gave rise to the word 'finch', originating from its call *fink! fink!* combined in its name with the term 'chaff', in the midst of which it could be seen searching for seeds on wheat threshing floors.
- Feeding flocks of Chaffinches that form after the breeding season frequently exhibit sexual segregation, consisting of either male or female birds.
- Studies have discovered that Chaffinch songs are not identical across Britain, but instead have local variations, or 'regional dialects'.

## Not quite a Chaffinch?

Female Chaffinches may be confused with female House Sparrows, which lack the prominent white wing markings (*see House Sparrow*), while male Chaffinches could be muddled with male Bullfinches, which have a black cap and much brighter red chest (*see Bullfinch*). But if your Chaffinch looks a little more orange than usual and has a white rump, which is most visible in flight, then you could be looking at its northern cousin: the Brambling. These birds from Scandinavia visit Britain in variable numbers during the colder months in search of milder weather and seeds, in particular beechmast. Bramblings may join flocks of Chaffinches and visit gardens, but return north in the spring. Enjoy the sightings before they leave.

Brambling

# Bullfinch

Female

## Key identification features

- The Bullfinch is a plump, bull-necked, sparrow-sized finch.
- The male is one of our most handsome birds – his plumage a smart and simple palette of reddish-pink front, grey back, black wings and tail.
- Both males and the plum-brown females, have a glossy black cap which extends forwards around the stubby little dark beak, making it seem almost non-existent. They also have white rumps which are particularly eye-catching in flight.

## Where could you find one?

The Bullfinch can be found year-round throughout most of Britain in suitable habitat. It feeds on seeds, fruits and berries, among other things, so needs to live where there are plenty of trees, weeds and hedgerows and is found in mature gardens, woodland, lowland farmland and orchards (where an appetite for fruit buds means it can be considered a pest). They will visit bird tables for sunflower seeds and other snacks.

## Is it easy to identify?

You would think so, especially given the male's red front. The problem is that Bullfinches don't tend to show off their good looks. Instead they keep a low profile, discreetly going about their business out of view in hedges and among tree branches. Invariably in pairs, they keep in touch with one another in cover with soft, sad, simple whistles – which may be the only clue to their presence. But even just a glimpse of a Bullfinch can be enough to identify them: the stocky shape, black cap, small bill and either rosy-red front on the male or toned-down pinkish-brown hues of the female. In flight the white rear end, a bright square between grey back and black tail, is the giveaway.

## How likely am I to see one?

Handsome male

This largely sedentary species is widespread and common and currently numbers around 190,000 pairs in Britain. A good view can be hard to come by, given they like to keep themselves to themselves, but once you spot one you should typically see its other half close by, or even a family group. They will visit bird tables and in the winter a rosy male perched on a frost-coated branch is every nature photographer's dream.

## What makes it special?

Sensitive, modest, devoted and conscientious are not the kinds of terms one usually associates with birds, but the Bullfinch is a unique species that deserves a little anthropomorphic indulgence. While other finches can be noisy, showy and sociable, the Bullfinch is quite the opposite. For a start, males and females are seemingly inseparable – our own little lovebirds. You almost always find pairs together, foraging within calling distance, whatever the time of year. The males don't flaunt their dapper looks from prominent perches, or sing like other birds to stake their claim to a territory. In fact, their song (what little of it there is) is a discreet refrain only intended for the ears of a mate. Surprising then, that this generally quiet species has a secret vocal talent: an uncanny flair for musical mimicry. The ability of Bullfinches to learn complex whistled tunes, diligently practising and perfecting them, made them a very popular cage-bird in Victorian times, coupled with their affectionate, if delicate, nature.

## Added interest

- Bullfinches were widely persecuted because of their taste for the buds of fruit trees, which they can scissor off at the rate of thirty a minute. However, they only tend to raid orchards when other food supplies, such as ash seeds, are in short supply.
- Parent birds use pouches in the bottom of their mouths to store food when foraging for their young – a special adaptation that reduces the number of trips to and from the nest.
- It is believed Bullfinches form faithful pairings for life and the male's exceptionally small testes are considered strong evidence of a monogamous lifestyle.

## Not quite a Bullfinch?

If your Bullfinch is missing its black cap then it could be a juvenile. However, if you spot a large red finch sitting at the top of a conifer like a Christmas decoration then you might have got lucky – it could be a male Crossbill. The mandibles of these bizarre-looking birds cross over at the tip, enabling them to extract seeds from pine cones.

There are approximately 40,000 pairs of common Crossbills in Britain, but local numbers tend to fluctuate as they move around. The males vary in the intensity of their red colouring, while females are greenish. Crossbills hang out high up in the trees and only come to ground to drink, so are not always obvious.

Crossbill

# Goldfinch

Juvenile

## Key identification features

- The Goldfinch is a colourful, slim little finch that looks like it has been overdoing the make-up. It has a black crown, white cheeks, a red face and is an attractive sandy-brown on the back and in two patches on either side of the chest.
- It has black and white in the tail and dark wings which have a yellow band running down the centre – a bar of buttery brightness that is noticeable when flying and as a patch on the folded wing. Juveniles lack the face markings, but still have the yellow in the wing.

## Where could you find one?

Mostly concentrated in the southern half of Britain and absent from the far north of Scotland, Goldfinches are common resident birds of scruffy open land where there are plenty of seed-bearing weeds. Nesting in trees or bushes, they can be found along farmland field edges and hedgerows, in wasteland, on road verges, in orchards, churchyards, parks, woodland clearings and also increasingly in gardens, where they will visit bird feeders.

## Is it easy to identify?

With its clown-like face paint and yellow flash along the wing, the Goldfinch is a distinctive-looking little finch – especially if you get a decent view, which can be had when they are busy feeding. Close up, the red patch covering the forehead and around the base of the beak looks like they have been messily tucking into a pot of strawberry jam. They are light and agile enough to cling on to the flower heads of thistle and teasel as they extract the seeds (in Anglo-Saxon times they were given the wonderful name 'thistle-tweaker'). Outside the spring breeding season, small feeding flocks can be seen flitting about with a light and skipping flight on weedy patches of land, twittering as they go. At a distance they are best identified by the broad brush-stroke of bright yellow on the wing.

## How likely am I to see one?

While absent from the far north, Goldfinches are common birds (around 1.2 million pairs breed in Britain), widespread across a range of habitats and are especially noticeable in the late summer. Numbers drop off in the winter as many head south to the Continent to avoid freezing weather before returning the following spring.

Attractive plumage

## What makes it special?

Few birds seem quite as cheerful as the Goldfinch. Flocks of these attractive yellow-winged little rays of sunshine never fail to delight. There is something irrepressibly joyful about their pretty plumage, their liquid twittering, gregarious nature and bouncy flight. However, their appeal was their undoing in the nineteenth century as large numbers were trapped for the cage bird trade, causing local populations to plummet. Their fate became an early campaign cause for the then fledgling RSPB, which was set up in 1889, and numbers bounced back. In recent years, this weed seed-addict has found a friend in the garden bird enthusiast and more and more are now visiting feeders for free handouts of their favourite nyjer seeds and sunflower hearts. For those who think British birds are brown and boring, just point out a Goldfinch, with its tawny breast, yellow wing feathers and head patterned in flag-like blocks of black, white and red.

## Added interest

- The collective noun for a group of Goldfinches is a 'charm' – a term with origins in the tinkling chorus of a flock, rather than its appearance.
- Female Goldfinches are more likely to hop across the Channel and spend the winter in France and Spain than males, who tend to sit tight, closer to breeding territories.
- Male Goldfinches have slightly longer beaks than females, enabling them to more easily extract seeds from the heads of teasel.

## Not quite a Goldfinch?

If it is summer and the finch you are looking at lacks the red, white and black face markings, then it could be a juvenile – young birds will moult into their full plumage in the autumn. If your plain-faced finch looks quite heavyweight, with a noticeably thick seed-cracking bill and has yellow in the wing (like a Goldfinch) and in the tail (unlike a Goldfinch), then it could be a Greenfinch. Male Greenfinches are olive-green, but dull-brown females and young birds could be confused with juvenile Goldfinches (*see Greenfinch*). Finally, if the streaky little finch you are watching feeding acrobatically in a tree has yellow in the wings, tail and on the sides of the face, then it could well be a Siskin (*see Siskin*).

Juvenile Greenfinch

# Greenfinch

Female

## Key identification features

- The Greenfinch is a stocky finch with a pale and noticeably hefty beak, capable of cracking large seeds.
- The male is olive-green with yellow at the sides of the short, slightly forked tail and a flash of yellow in the wing, visible as a vivid streak along the edge when it is folded.
- The female and juvenile resemble a duller brown version of the male. During the winter, male Greenfinches also look much less colourful, but by spring their plumage is at its brightest best.

## Where could you find one?

This dedicated seed-eater is found across Britain year-round in suitable areas with trees, weeds and shrubs, from lowland woodlands and farmland hedgerows to orchards, churchyards, villages, urban parks and gardens. Greenfinches are less common than they once were in farmland areas, where spilt grain and weed seeds are harder to find, but have become dedicated bird table visitors with an appetite for sunflower seeds and peanuts among other treats.

## Is it easy to identify?

In overcast conditions the plumage of the male Greenfinch can look a bit mossy-grey and lacklustre, but in the right light he is a wonderful cooking apple green and the yellow along the edge of the folded wing stands out as a striking streak of lemon brightness. This yellow wing feathering is a key feature in less-colourful females and juveniles as well. The Goldfinch and Siskin also have yellow in the wing, but they are more delicate-looking and lack the fearsome thick beak of the Greenfinch. In flight, the Greenfinch not only shows the spray of yellow across its grey outer wings, but also distinctive wedges of yellow on the sides of the dark tail. During the spring and summer breeding season they are at their most vocal and can be identified by their twittering songs which contain a tell-tale nasal wheezing *dweee* sound.

## How likely am I to see one?

The Greenfinch is a common resident species, with an estimated 1.7 million pairs across Britain wherever there are plenty of trees, bushes, shrubs and weeds. They can be seen on a woodland walk, nature reserve ramble, parkland stroll, along hedgerows and verges or in the garden and often draw attention to themselves with their calls. Outside the breeding season they will also join together with other finches and seed-eaters in feeding flocks.

A large flock of feeding Greenfinches

## What makes it special?

At one time a country-dweller, the Greenfinch has lost some of its wariness and embraced a more suburban existence in order to take advantage of bird table handouts, enabling us to enjoy decent views of these attractive birds. A handsome male is a stunning sight during the spring, when his green body and yellow blaze on wing and tail are at their boldest. The combination of smart lemon-and-lime looks and pleasant twittering song, add colour and cheer to any garden, even if they can be bossy at the bird feeders. Their burly physique, frowning brows and sturdy beak are enough to frighten off smaller birds as they bicker over the peanuts and sunflower seeds. Tough they may appear, but of all the finches they have suffered the most in recent years from a fatal parasitic disease transmitted among birds, with outbreaks reducing local populations by up to a third. Regularly cleaning bird feeding stations is recommended to help prevent the parasite being spread.

## Added interest

- There may be many more Greenfinches visiting your bird table during winter than you imagined. In one study over 1,000 individuals were ringed in a garden over two months, even though only a handful were ever seen at a time.
- The Greenfinch's stout, conical bill enables it to tackle a wide range of food, from large rosehips to sunflower seeds and peanuts.
- In the spring male Greenfinches perform flight displays, twittering and circling at tree-top height with slow exaggerated wingbeats.

## Not quite a Greenfinch?

If your Greenfinch is on the small side, but greenish-yellow nonetheless, it could be a Siskin - a streaky little finch found in conifer forests or feeding acrobatically among alder and birch trees. This delightful bird, abundant in Scotland and Wales, is always a treat to spot (*see Siskin*). If there is yellow along the edge of the folded wing, sandwiched between dark feathers, with clown-like black, white and red face markings then this is a Goldfinch (*see Goldfinch*). Finally, if you are in coniferous woodland and you spot a solid-looking greenish-grey finch, examine the beak. Should both halves be twisted over at the tip, as if it has been in a rear-end shunt, then this is a Crossbill - an exciting find (*see Bullfinch*).

Siskin

# Siskin

Female

## Key identification features

■ The Siskin is an attractive, agile, little yellow-and-black finch. Smaller than a House Sparrow it has a pointed beak, ideal for extracting seeds from pine cones and is able to hang upside down when feeding.

■ The eye-catching male is bright yellowy-green on the chest and face, with a yellow bar on his dark wings and some yellow in the tail. He has a black cap and a little bib of black under his bill and streaky flanks. Female and juvenile birds are far plainer and more streaky.

## Where could you find one?

The Siskin is mainly a bird of coniferous forests, where it feeds on spruce and pine seeds and is most abundant in Scotland and Wales during the spring and summer breeding season. Flocks roam widely in search of tree seeds in mixed and evergreen wooded areas and, during the winter, Siskins can be found across the UK – and are an increasingly common sight at garden bird feeders.

## Is it easy to identify?

Siskins tend to feed high up in trees, so may be spotted as silhouettes: small, chatty, restless birds hanging upside down, like members of the tit family, probing their bills into cones in search of seeds. They have a noticeable fork in the tail and have a clear, slightly questioning call: *seeuuee*. As they are invariably seen in groups, it is easiest to identify the species by the brighter males, with their black caps and chins. A good view reveals the male's attractive yellow plumage, which contrasts with dark feathering: in particular, a conspicuous stripe of yellow on the side of the folded wing sandwiched between wedges of black. As they flit between trees, or to and from a bird feeder, splashes of yellow are visible in both the tail and along the dark wings.

## How likely am I to see one?

An acrobatic feeder

It is a lot easier to see Siskins than it once was - a handful of seed mix on the bird table anywhere in Britain could do the trick. They have increased in number and range thanks to plantation conifer forestry and free handouts at bird feeders. There are more than 400,000 breeding pairs and they are most widely spread across Britain in winter.

## What makes it special?

What Siskins lack in size they make up for in character. These dynamic little finches are a welcome sight in conifer forests and are even more of a delight when they pay our gardens a visit. Canary-sized, with yellow colouring, gymnastic agility and lively chatter, they can look every bit like imported cage birds that have escaped confinement and ended up among the usual suspects squabbling over the peanuts and sunflower seeds. Although they are becoming more familiar bird table visitors, they still feel like a novel addition to our garden fauna. Learning to take advantage of our generosity through the hard winter months has undoubtedly helped this small species to prosper over the last half century. However, we have also provided food for Siskins on a far larger scale - through plantation forestry. Maturing stands of seed-bearing non-native conifers prove an ideal habitat in which to feed and raise young.

## Added interest

- It wasn't until the 1960s that Siskins first started visiting bird tables and, initially, it was peanuts hung up in red mesh bags that attracted them, though they now eat from all kinds of feeders.
- Siskins rely on a number of tree seeds to get them through the year, plundering the cones of coniferous spruce and pine during the breeding season and feeding on birch and alder during the autumn and winter.
- They are social birds with group hierarchies and dominant birds may be fed by those lower in the rankings.

## Not quite a Siskin?

There is one exciting find that frequently mixes with Siskins: the Lesser Redpoll (or commonly 'Redpoll' for short). Brown and streaky with a forked tail, this finch shares the habit of feeding in flocks hanging upside-down. However, it lacks any hint of yellow in the plumage and has a charming smudge of dark red on the forehead. Lesser Redpolls are widespread birds of birch thickets and woodlands, particularly in northern and western areas during the breeding season where they number over 200,000 pairs, but they have suffered serious long-term declines. If your Siskin has black, white and red on the head then it is a Goldfinch (*see Goldfinch*). If it looks on the large side then it could be a Greenfinch (*see Greenfinch*).

Lesser Redpoll

# Blackcap

Female

##  Key identification features

■ The Blackcap is a grey-brown bird, almost Sparrow-sized, with a distinctive cap on top of the head: black for males and chestnut-brown for females.

■ Apart from the cap, they are quite plain and lack any eye-catching plumage colours or white feathers in the wings or tail. They are slightly darker grey-brown on the back and paler beneath and have a fine dark bill.

##  Where could you find one?

The Blackcap is, for the most part, a summer visitor, found across the UK but more abundant in the southern half of Britain. It breeds in woodlands, bushy copses and mature gardens and parks – anywhere leafy with plenty of undergrowth that offers a little privacy, plenty of insect life and dense tangles of foliage in which to build its nest. Those that spend the winter here regularly visit bird tables.

##  Is it easy to identify?

The clue's in the name. Without the black cap this would be a rather nondescript species, but the neatly defined inky crown, running to eye level from front to back, makes it one of the easiest warblers to identify – and they're a tricky bunch at the best of times. Not only that, but it also helps distinguish the sexes of adult birds, as the females have a reddish-brown cap. When they are feeding discreetly out of view among tree foliage they can be hard to spot, but a glimpse of the cap atop a plain grey head is enough to identify them. Their colour is in their voice. In the spring, the territorial males draw attention to themselves with florid outbursts of rich fluty song.

## How likely am I to see one?

Territorial singer

Over a million pairs of Blackcaps breed in Britain, spread out across the country in suitable wooded and well-vegetated habitat. In the spring, they are easiest to spot before the trees are in full leaf and you may hear one first – a burst of splendid song that helps locate the territorial male singer. Our Blackcaps head south in the autumn and are replaced by several thousand from Europe that spend the winter here.

## What makes it special?

The Blackcap is one of our finest spring singers and its complicated melodies have earned it the nickname the 'northern Nightingale'. Male Blackcaps sing to attract mates and defend territories and their engaging outpourings of melodic warbling contain both rapid scratchy sounds and lush Blackbird-like whistles. In the autumn, once the business of raising young is over, Blackcaps depart our shores and migrate south through France to enjoy a little winter sun in Spain, Portugal and north and west Africa.

Over the last half-century, increasing numbers have been recorded here during the colder months and bird ringing studies have now revealed that they are actually individuals from central Europe whose genetic sat-nav sent them in the wrong direction – west instead of south. These muddled migrants should have perished during Britain's chilly winters, but garden bird feeders in warmer urban areas provided a lifeline and have fuelled the evolution of a distinct group of short-haul pioneers.

## Added interest

- Blackcaps are bossy birds at garden feeders, monopolising the suet and seeds by frightening away smaller species. An additional feeder some distance away can get around the problem.
- Blackcaps have a taste for sweet foods such as berries, windfall fruit and also, perhaps surprisingly, nectar. A pollen-dusted face may indicate they have been feeding from flowers.
- Only adult males have black caps – juvenile birds of both sexes look similar to brown-capped females until they moult into their grown-up breeding plumage.

## Not quite a Blackcap?

If your Blackcap is on the small side, rotund and has a little black chin and white cheeks it could be a Marsh Tit (*see Coal Tit*). Missing the black cap altogether? Then it might be the closely-related Garden Warbler. Numbering 170,000 pairs in the spring and summer, the Garden Warbler has an enchanting bubbling warbling – very similar to a Blackcap. Finally, if your similar-sized singing warbler has a grey top half to the head, rather than a black cap, it could be a Whitethroat. The 'Lesser Whitethroat' is the most similar, while the more common and widespread 'Whitethroat' has a brown back and both have white throats. The Lesser Whitethroat numbers 74,000 pairs while the Whitethroat tops a million pairs.

Whitethroat

Juvenile

## Key identification features

- The Chiffchaff is an active little brown bird: delicate, slim and smaller than a Sparrow, with a thin beak and dark legs.
- It has a dull greyish-olive back and is lighter underneath, with a hint of buff-yellow on the breast and throat. It lacks splashes of colour, bright stripes, white feathers or any stand-out plumage.
- The only mark of note is a light eyebrow – though this is also present on a number of other similar warblers, including the virtually identical Willow Warbler. The Chiffchaff's voice provides a key to identification (*see below*).

## Where could you find one?

This summer visitor can be found in lowland areas across most of Britain, though less so in the far north. It is an insect-eater that breeds in deciduous woods with plenty of undergrowth, parks, bushy places and occasionally visits gardens. Chiffchaffs arrive in the spring and depart in the autumn, heading south to the Mediterranean and northern Africa, though a small number spend the winter here.

## Is it easy to identify?

The Chiffchaff is about as nondescript a warbler as you can get and difficult to view clearly as it is constantly on the move, occasionally twitching its tail in a restless fashion. It is also extremely hard to tell apart from its close relative the Willow Warbler, which has pinker legs – observers in doubt simply refer to them by the term 'Willowchiffs'. That is until they sing, at which point you no longer need to be an expert to know what you are looking at. The Chiffchaff pronounces its name loud and clear: *chiff-chaff, chiff-chaff, chiff-chaff*. The Willow Warbler's song is a short and fluid cascade of sweet descending notes. Listen to a birdsong CD or online and you will soon be able to impress those around you by pausing on a spring walk, cupping an ear and declaring 'ah, a Chiffchaff', before pointing out an obscure-looking little bird in a distant tree.

## How likely am I to see one?

A strident songster

You are most likely to hear a Chiffchaff before you see one, as its strident, repetitive song engages the ear more than its understated plumage catches the eye. Spring is the time to listen and you should get lucky as the Chiffchaff is a common summer visitor, with over a million pairs coming to Britain to breed. They fly south in the autumn and only a thousand or so are recorded here during the winter months.

## What makes it special?

The Chiffchaff is one of the earliest summer migrants to arrive in Britain, touching down in March in advance of the great spring invasion of Swallows, Swifts, Flycatchers, Cuckoos and the rest. Its simple song, delivered with vigour from the bare branch of a tree not yet in leaf, lifts the spirits, heralding longer, warmer days ahead. And when it comes to identification, it helps that it has been given such an accurate onomatopoeic name based on its rhythmic ditty, *chiff-chaff, chiff-chaff, chiff-chaff*. Learning birds from their sounds can feel like something for the experts, but with this common species it is far easier than relying solely on visual clues, as it looks so similar to other warblers. Its closest match is the Willow Warbler, a longer-distance migrant which is twice as abundant and lives in similar habitat but has a far more musical song that runs down the scale in a delightful fashion.

## Added interest

- The Chiffchaff looks so similar to the Willow Warbler that it took until the eighteenth century before Hampshire curate and renowned naturalist Gilbert White pronounced them separate species.
- At one time the Chiffchaff was given the endearing name 'Lesser Pettychaps' based on its small size.
- Male Chiffchaffs focus their efforts on singing and defending a territory, leaving the females to build a nest and raise the young by themselves.

## Not quite a Chiffchaff?

Many warblers share the Chiffchaff's slim build, dull colours and thin little bill. The Willow Warbler has been mentioned above, while the Wood Warbler is a much yellower bird and a lovely find. This migrant is only in Britain during the spring and summer breeding season and numbers about 6,500 pairs. The Reed Warbler is a browner bird that has a population of 130,000 pairs, while the Sedge Warbler is twice as common and has a noticeable cream streak above the eye. Finally, if your Chiffchaff is sitting upright in a tree looking a bit grey-brown and has a streaky light underside then it could be a Spotted Flycatcher, a widespread summer migrant of woodland glades (*see Meadow Pipit*).

Willow Warbler

# Pied Wagtail

Female

## Key identification features

- The Pied Wagtail is a dainty, active, little black and white bird, a bit bigger than a Sparrow with an elongated tail which it almost constantly wags up and down.
- It has a thin dark bill, relatively long legs and walks, or runs, rather than hops.
- The male is black on top, white underneath and has a white face and black throat and chest, while the female is similar-looking but dark grey on top. In flight, white is visible in the black wings and along the outer edges of the slim black tail.

## Where could you find one?

Supermarket car parks, factory roofs, golf courses, railway station platforms, pavements… and out in the wild countryside as well! This versatile insect-eater likes flat areas where it can easily spot and chase after small invertebrate prey and isn't too fussy where – anything from tarmac or mud to close-cropped turf will do. It favours habitat near water, including the edges of reservoirs and gravel pit lakes. Pied Wagtails are common and scattered throughout Britain. In winter they retreat from northern and upland areas and flocks can be seen roosting in city centres.

## Is it easy to identify?

Fortunately its straightforward name tells the story: it is pied and it wags its tail. This slender black and white bird is quite relaxed around people, so can be viewed with comparative ease. It is a restless, busy little species that dashes from one spot to another, picking up insects or launching itself into the air after them. The head markings are distinctive, especially the patches of white around the dark eyes and the tail seems to have a life of its own, relentlessly twitching up and down like a faulty speedometer needle. The Pied Wagtail also has a distinctive call, often delivered in flight, which sounds a bit like the name of the west London borough *Chiswick*!

## How likely am I to see one?

A roosting flock

This is a widespread species, numbering over 450,000 pairs in Britain. In summer they are distributed throughout the country, during colder months they tend to be found in more southerly areas, gathering in flocks to roost together in woods, reedbeds, trees and flat roofs at service stations and supermarkets. They will visit gardens and may raise young in open-front nest boxes.

## What makes it special?

Despite being a common enough sight, the Pied Wagtail is not as widely known as other everyday birds. Somehow this delicate and sprightly species gets overlooked, despite constantly flagging up its presence with a waving tail. They energetically go about their business in full view, hardly attracting a glance from passers-by as they search for food. But they are certainly deserving of admiration. Pied Wagtails have an unflagging zest for life: skipping, darting and dancing their way through it. Not only are they attractive and balletic birds, they also bring cheer to the most dismal of man-made settings. They animate barren stretches of tarmac and roost in the kinds of ornamental urban lollipop trees that would otherwise be devoid of wildlife. Autumn and winter roosts can be an impressive spectacle, as dozens or even hundreds gather together for the night, seeking out warmth and safety in numbers, sometimes right in the heart of cities.

## Added interest

- The Pied Wagtail is, in fact, a British version of the Continental White Wagtail, which is light grey on the back and can be spotted passing through the UK on migration.
- The Pied Wagtail was given a variety of nicknames in the past that play on its waterside haunts and pumping, dipping, tail action, including Nanny Washtail and Polly Washdish.
- Car parks are popular invertebrate hunting grounds for Pied Wagtails, which have learnt to pluck insects from car radiators, bumpers and tyre treads.

## Not quite a Pied Wagtail?

If your Pied Wagtail has a grey back and is lemon yellow underneath, then it could be a Grey Wagtail. These resident birds also wag their tails and are found near water – though particularly fast-flowing rocky streams and rivers. Widespread, but less common than the Pied Wagtail, they have a yellowish rump, visible in bounding flight as they make off along watercourses, alighting on boulders mid-current with long tail pumping (*see Grey Wagtail*). The third of our wagtails and the scarcest, is the Yellow Wagtail, a summer visitor to damp pasture, meadows and arable fields, mostly in England. It has a greenish back, yellow in the face and underneath, and is a lovely bird to spot, though sadly in decline (*see Grey Wagtail*).

Female Grey Wagtail

# Song Thrush

## Key identification features

- The Song Thrush is a smart, well-proportioned bird, a little smaller than a Blackbird. It sings loud and clear from a prominent perch, repeating phrases.
- Both sexes are warm brown on top and light below and flushed buff-yellow on the chest.
- The breast, flanks and underside are speckled with black dots that are shaped like arrowheads and the legs are pale and pinkish. The tail and wings are plain brown, without white feathers and the 'armpit' area, visible under the wing in flight, is orangey.

## Where could you find one?

Common and widespread, the Song Thrush is distributed across Britain in suitable habitat which combines open areas for feeding close to trees and the cover of bushes. As such it can be found year-round in parks, woodland, farmland and along hedgerows. Gardens with lawns and shrubs are ideal and the Song Thrush will feed on scattered fruit, mealworms and other treats under bird tables.

## Is it easy to identify?

A breast peppered with dark spots and brown back make this garden regular fairly distinctive, as it hops and runs across a lawn in a stop-start fashion looking for worms. However, it does have a close relative that is very similar: the Mistle Thrush. This is a bigger, tougher, paler bird whose back is a greyer brown and the dark dots on its creamy underside are rounder. In flight its 'armpit' is white, as opposed to orangey in the Song Thrush. The Mistle Thrush also has a harsh chattering alarm call, which has been described as sounding like an old-fashioned football rattle and a melancholic and fluty song. The more inventive and musical Song Thrush delivers a series of short, unpredictable phrases, ranging from a couple of simple whistles to complex rapid warbles and chosen fragments are repeated several times.

## How likely am I to see one?

The Song Thrush is a relatively common species, with over a million pairs scattered across Britain. They are territorial birds during the spring and summer breeding season and some remain on their home patch over the winter, while others may move south - even as far as France and Spain.

A fine singer

## What makes it special?

The Song Thrush is one of our finest singers, with an incredible ability to incorporate dozens of intricate phrases (even mimicry), into its musical compositions. Its exquisite refrains are sung clearly and loudly, often filling the air at quiet times of the day. If you are lucky enough to have one of these songsters in the garden it is worth taking the time to marvel at its creativeness. Some of its short phrases are harsh and scratchy, others rich and pure and it delivers its sequence of variations one after another, reiterating for emphasis and pausing for effect, leaving you in no doubt that you are witnessing a performance. However, the song is heard far less often than it once was, as the population has more than halved since the 1970s and it has now been placed on the UK's Red List (*see page 186*). The Song Thrush is still common enough to be a garden regular and numbers appear to have turned a corner in recent years. Savour the sightings - and enjoy the free concerts.

## Added interest

- The Song Thrush eats garden snails by smashing their shells on hard surfaces and regularly-used 'anvils' are typically littered with shell fragments.
- Male Song Thrushes sing in order to proclaim their territory and attract a female - and they do so a lot less in the spring once they have secured a mate.
- Females are hard-working and dedicated mothers, laying up to three clutches a year and even building the cup-shaped nests, which they line with mud, dung or decayed wood.

## Not quite a Song Thrush?

If your Song Thrush is on the large side and greyer looking, then it could be a Mistle Thrush (*see Is it easy to identify? section*). Song Thrushes tend to feed near cover, while Mistle Thrushes are perfectly at home out in the open, have rounder spots on the underside, white underwings and a harsh rattling call and have a population of 160,000 pairs in Britain. Song Thrushes may also be muddled with young or female Blackbirds - mottled fronts but are a darker chocolate brown all over, or Redwings - a white eyebrow and rusty-red flanks (*see Fieldfare*) or, on open grassland you might spot a dainty little bird with a thrush-like speckled breast, which could be a Meadow Pipit (*see Meadow Pipit*).

Mistle Thrush

# Blackbird

## Key identification features

- The Blackbird is a familiar member of the thrush family, with unfussy plumage and a simple body shape.
- The handsome male is most recognisable, being coal-black all over apart from an orangey-yellow beak and, on close inspection, a thin yellow ring around the eye.
- The female is dark brown, slightly lighter under the throat and mottled on the breast. Young birds look like females, though are more speckled all over and have a dark beak.

## Where could you find one?

Common and widespread across Britain all year round, Blackbirds are found in a range of habitats including woodland, scrubland, farmland, parks and gardens. Pretty much any lawn is likely to have a Blackbird hopping about on it in a stop-start fashion, head cocked, looking and listening for worms. Gardens provide ideal habitat, combining open areas of close-cropped turf with the protective cover of trees, shrubs and hedges, within which they build their cup-shaped nests of grass and mud.

## Is it easy to identify?

The male keeps things straightforward when it comes to identification, being a black bird. The glossy plumage is uniform in its darkness and his only concessions to colour are a bright orangey-yellow bill and narrow circles of yellow around the eyes. The female is chocolate brown, with less yellow in the bill and some speckling on the front. Juvenile Blackbirds look much like females – young males take months to develop their black plumage and yellow bill. Blackbirds are noisy, drawing attention to themselves as they rustle through leaf-litter looking for invertebrates and with their varied, sometimes overly-agitated, alarm calls. Territorial males tend to sing from lofty perches and make a show of landing on a branch or patch of lawn, cocking their tails with swashbuckling swagger. At dusk the persistent *chink, chink, chink* sound that can be heard echoing through suburbia comes from restless Blackbirds settling down for the night.

## How likely am I to see one?

A sunbathing male

The Blackbird is one of our most abundant species, numbering five million pairs in Britain, increasing to 10-15 million birds in the winter as migrant Blackbirds from northern Europe arrive to see out the colder months. It is also one of the species most commonly spotted by those taking part in the annual RSPB Big Garden Birdwatch.

## What makes it special?

It is easy to take such a common garden bird for granted, but a male Blackbird is an exceedingly handsome bird of simple sophistication: smartly suited in black with a distinguished yellow bill. He is also one of our finest avian vocalists, signposting his territory with languid, rich and melodious song, each verse distinguished from the next by a brief pause during which he can listen for neighbouring singing males. There is a hint of melancholy in the tender croakiness of his voice and descending notes. The mellow fluting refrains differ from both those of the Song Thrush, which repeats its varied phrases and the strident, desolate passages of the Mistle Thrush (*see Song Thrush*). While the male Blackbird's song may sound effortless, it is deceptively complex. He is a skilful composer, embellishing his repertoire throughout the few years of his life with fresh variations to impress potential mates.

## Added interest

- Young Blackbirds clamber out of the nest before they are able to fly, but continue to be fed by parent birds until they can take to the wing. They are not as helpless as they appear, so avoid the temptation to 'rescue' feathered fledglings.
- The colour of a male's beak is a barometer of vitality. If the bird is in good condition, its immune system is able to free up the pigment which turns the yellow bill a healthy shade of golden-orange.
- In sunny weather, Blackbirds engage in a spot of sunbathing on the ground to help rid their plumage of parasites, appearing blissfully relaxed with wings spread, feathers ruffled and beak open.

## Not quite a Blackbird?

If your Blackbird has a white feather or two, it may be exhibiting a condition known as leucism - a partial loss of pigment which may be inherited. A rare individual that is completely white must have pink eyes to be a true albino. If you happen to spot a Blackbird in an upland area with a white bib, you may have struck lucky - it could be a Ring Ouzel. The name 'ouzel' was the Old English word for a Blackbird and the 'ring' refers to the broad band of white across the chest. About 6,500 pairs arrive in the spring and breed in mountains and moorlands in north and west Britain before migrating south for the winter. An excellent sighting!

Ring Ouzel

# Starling

Breeding plumage

## Key identification features

- The Starling is a common, confident bird with dark plumage and a pointed beak.
- In the spring and early summer the bill is yellow, while its feathers are black and have an iridescent purple-green sheen and some pale dots.
- In the autumn and winter it takes on a speckled look, with light spots peppering its front and back and the bill turns black.
- On the ground, it walks rather than hops and, in flight, has a noticeably short square tail and triangular wings.

## Where could you find one?

The Starling can be found year-round pretty much everywhere across Britain, apart from upland areas. It nests in holes and cavities in buildings, cliffs and trees; feeds on farmland, playing fields, in gardens and along shorelines; and roosts in woods, reedbeds and on industrial sites. This is a gregarious and resourceful species of both urban and rural areas.

## Is it easy to identify?

The Starling is fairly straightforward to identify, though could perhaps be confused with a Blackbird which is also dark and has a yellow bill. Starlings are smaller though, with a shorter tail, have an oily iridescence and are attractively freckled with white spots in the autumn and winter. Their pointed appearance, stocky build, slick-gelled sheen and jerky stride give them a certain street-wise demeanour. Greedy at feeders and messy at roosts, they are not popular with all, but in the right light they have a glorious metallic shine to their plumage and are full of character. Telling the sexes apart is tricky, but possible with close observation of the yellow bill: in males the base is blueish, but light pinkish in females. Juvenile birds are muddy-brown and gradually acquire spots on the body, then on the head, after moulting.

## How likely am I to see one?

Starling chicks in their nest

The Starling is one of the most commonly recorded birds in the annual RSPB Big Garden Birdwatch. Its breeding population totals nearly two million pairs and numbers soar in the winter when our resident Starlings are joined by birds from northern Europe. In the spring, Starlings nest in convenient cavities, such as under roofs or in old trees and will use hole-fronted nest boxes, laying exquisite pale blue eggs.

## What makes it special?

The Starling's song is a bit of a mechanical medley of jangling, rattling, wheezing and whistling sounds. Not the most refined of singers, but they are incredible mimics – from the calls of buzzards to car alarms. Captive Starlings can even be taught to imitate words or learn tunes and one particularly musical individual was kept as a pet by Mozart. However, it is in numbers that they are most impressive of all. In the colder months, groups of birds gather in vast flocks to roost in the evening. The sight of tens, or hundreds of thousands of Starlings swirling in dense clouds against a clear winter's sky is one of Britain's great wildlife spectacles. Known as a 'mumuration', the pulsing congregation can resemble a single organism, stretching and twisting as one before finally settling for the night. Unfortunately these aerial displays are becoming a rarer sight, as many of the vast roosts have dwindled. While still abundant, Starling numbers have declined significantly over recent decades and they have been included on the UK's Red List.

## Added interest

- Male Starlings embellish their crude nest-building attempts with aromatic green foliage, which may help ward off parasites and impress females.
- When probing the ground in search of invertebrates, Starlings are equipped with strong muscles that enable them to force open their bills within the soil to see what lurks beneath.
- Females that have no permanent partner may lay eggs in other Starling nests, waiting until an unwitting pair are preoccupied elsewhere before sneaking in, Cuckoo-style, to add to their clutch.

## Not quite a Starling?

If your Starling looks the right shape and size, but lacks a glossy sheen or any white spotting, it could be an immature bird. These dull grey-brown juveniles, often seen in groups in late summer and autumn, gradually develop spots on the body and eventually the head. Blackbirds may also be confused with Starlings, but are a uniform matt black, more elongated and not flecked with white spots in the winter (*see Blackbird*).

Blackbird

Juvenile Starling

# Woodpigeon

## Key identification features

- The Woodpigeon is the largest and most common of our doves and pigeons.
- It has a smooth blue-grey body, a prominent white patch on either side of its neck and the uppersides of its wings have a bold white stripe running across the middle.
- Above the neck patch the feathers have an iridescent greenish-purple sheen and a black band runs across the end of the fanned tail. The eye is yellow and the plump breast has a pinkish hue. The sexes are alike, but young birds lack the white neck patch.

## Where could you find one?

Arable farmland, where Woodpigeon can feed on grain and vegetation, coupled with woodland copses in which to nest, provides the ideal habitat. They are most abundant in central, southern and eastern England, though are common all over the UK apart from northern Scotland and exposed uplands where life would be a little too demanding for this comfort-seeking species. In recent decades they have moved into urban settings and become garden regulars.

## Is it easy to identify?

The barrel-chested portly frame of a Woodpigeon is exaggerated by a small head and relatively short legs, which one imagines might buckle under their weight as they waddle across the ground. They can appear clumsy, smashing their way skywards when alarmed with a clattering of wings and scattering twigs and foliage. Yet they are good fliers, swift and strong and it is then that they reveal their key distinguishing feature: broad white crescents that run across the centre of each wing, between grey inner feathers and darker wing tips. The adult birds also have a paint-splodge of white on each side of the neck (which sets them apart from our other doves and pigeons) and close-up, their fronts are beautifully flushed with plum-pink. Their soft cooing song is a familiar spring and summer sound of town and country, and has a tell-tale quick-slow-slow-quick-quick rhythm, that could be described as: *I cooo for you-u*.

## How likely am I to see one?

Strong fliers

With over five million pairs in Britain, this year-round resident is hard to avoid. In the autumn and winter, numbers are boosted by all the young born during the summer, as well as by visiting birds from northern Europe, and large flocks form in arable areas. Woodpigeons build their flimsy twig nests in trees and tall hedges.

## What makes it special?

Woodpigeons are now among our most common birds. Their secret is that they eat their greens. Greedy feeders on grain, which has made them something of an agricultural pest, they also gorge on the buds, shoots and leaves of tender crops. Changing farming practices, including the increased cultivation of oilseed rape plants, means flocks now have plenty to keep them going through the hard winter months. In addition, Woodpigeons are able to raise young at any time of the year if the conditions are favourable. While birds don't generally applaud their own flying ability, a male Woodpigeon does make a show of things when engaged in territorial displays. After rising steeply into the air, he throws in a few almost self-congratulatory wing claps, before gliding downwards with tail fanned. This attention-grabbing display, often repeated in a roller-coaster type fashion, is a regular sight over parks and woodland edges – and impressive enough to deserve a little applause.

## Added interest

- Newly-hatched chicks are fed on a nutritious milky substance formed in the parents' gullet called 'crop milk' – an unusual behaviour also shared by some flamingos and penguins.
- While birds typically drink by scooping up water and tilting their beaks upwards to enable gravity to do the work, Woodpigeons are able to suck up water without lifting their heads.
- Woodpigeons are monogamous during the breeding season and reinforce the pair bond by engaging in the endearing practice of mutual preening.

## Not quite a Woodpigeon?

If your Woodpigeon is missing the white neck patches it is probably a young bird, but if it also lacks white braces across the open wings then it could be a Stock Dove. This common species, numbering 260,000 pairs in Britain, has glossy green neck feathering and dark edges to the wings. Another possibility is the street pigeon that comes in a hotchpotch of plumage variations and originates from the ancestral wild Rock Dove. Pure Rock Doves, which nest on cliff ledges, are found in north-west Scotland, have a white rump, twin black bars across the wings and are a satisfying bird to positively identify. Finally, grey pigeons in flight can be confused with falcons (*see Peregrine*).

Stock Dove

# Collared Dove

## Key identification features

- The Collared Dove is a slim, creamy, grey-buff bird with a telltale black half-collar on its neck.
- It is the archetypal elegant dove, both slender and attractive, with the faintest flush of pink in its plumage, a light biscuit-brown back and a pretty look about its face. Its wing tips are dark and a white outer edge to the longish tail is noticeable as it takes off and lands.
- Males and females look alike and the voice is a repetitive cooing, a familiar sound in suburban gardens where it is most at home.

## Where could you find one?

From rooftops to recreation grounds, Collared Doves can be found living alongside us across Britain wherever there is a source of food close at hand. They particularly favour gardens with trees, such as conifers, in which to nest or roost. Leafy suburbia is ideal, along with villages and farmsteads, so long as there is seed and grain to be found – whether scattered on lawns and open bird tables, in stubble fields or spilt around farmyards and agricultural stores.

## Is it easy to identify?

Juvenile

The distinctive collar, from which it gets its name, is the giveaway. Set against its pale and fairly uniform plumage, this black bracelet shape, emphasised by a thin white border, stands out even at a distance. However, immature birds lack this band and given they can be seen throughout the year these collarless Collared Doves might cause confusion. In general they are smart doves, with a certain grace that sets them apart from the scruffier street pigeon and the heavyweight Wood Pigeon. Their rhythmic three-syllable cooing, *coo-coooo-coo*, has become one of the quintessential sounds of suburban and country living. The bird they are most likely to be confused with is the more ornately patterned Turtle Dove which visits Britain during the spring and summer to breed (*see right*).

## How likely am I to see one?

Suburban regular

You don't have to go far to spot a Collared Dove - perhaps just to the kitchen window - as they live in close proximity to humans, inhabiting our leafier settlements. They regularly feature among the top ten species recorded by people in the annual Big Garden Birdwatch and currently number close to a million pairs across Britain.

## What makes it special?

Collared Doves are relatively recent UK arrivals, originating in the Middle East. Symbols of peace, the doves spread rapidly during the war years, conquering Europe and colonising Britain by the mid-1950s. In the six decades since, they have become no more surprising a sight in suburbia than the television aerials they perch on and the electric mowers which drown out their cooing on summer days. Whatever change of conditions or flick of a genetic switch caused collared doves to invade new territory suddenly, their progress was astonishing. From Turkey and the Balkans, the species' range expanded during the twentieth century across more than a million square miles, reaching as far as Scotland and Scandinavia. They are certainly smart, well-mannered additions to our wildlife, and their sandy-grey plumage and lightly-tanned backs hint at warmer climes and the hot and arid lands from which they hail.

## Added interest

- Collared Doves take a production-line approach to parenting, raising chicks in flimsy stick nests almost all year round when conditions are good - even feeding fledglings while incubating a fresh pair of eggs.
- While adults settle in an area, juveniles tend to roam in a north-westerly direction, which may help to explain how this successful Asian species eventually reached Britain in the mid-twentieth century.
- Like other pigeons and doves, they have a crop which enables them to store food. This throat pouch can also produce 'crop milk', a nutritious regurgitated substance fed to nestlings before they are able to digest seeds.

## Not quite a Collared Dove?

If your Collared Dove has all the standard characteristics but is missing a collar, then it is probably a young bird. But if it has a mottled marmalade-brown back and a more intricate black and white pattern on its neck, it could be a Turtle Dove. Perhaps the easiest way to remember the difference between the two species is that 'tortoiseshell' plumage means a 'turtle' dove. This attractive summer visitor has suffered an alarming decline over recent decades, due to habitat changes and over-hunting on migration routes, and has become an increasingly special sighting, even in its strongholds in southern and eastern England where it numbers fewer than 15,000 pairs.

Turtle Dove

# Great Spotted Woodpecker

## Key identification features

- The Great Spotted Woodpecker is a boldly marked, Blackbird-sized bird, that is every bit the woodpecker, with its pointed power-drill bill and vertical tree trunk climbing ability.
- It is black and white and has two stand-out features: a noticeable white patch on the shoulder and bright red feathering under its short tail.
- The sexes can be told apart as the male has a small square of red on the back of the head, which the female lacks. Juvenile birds have red on the front of the head.

## Where could you find one?

Mature woodland of any kind, whether broad-leafed, coniferous or mixed, can support Great Spotted Woodpeckers. They are found year-round in forests, copses, parks and large gardens across most of Britain, though in greater numbers in the southern half of the country, and will visit bird tables and hanging feeders.

## Is it easy to identify?

There is something instantly recognisable about a woodpecker: its formidable chisel bill and the way it clings to a tree trunk. We have three breeding species in Britain, but only two of them are pied: the small and scarce Lesser Spotted and the larger black and white Great Spotted, with its tell-tale blade of white running down the shoulder and a flash of crimson beneath the tail, as if it had been sitting in red paint. Surprisingly hard to spot when high up among the tree canopy, they draw attention to themselves by drumming on a branch or with their hard *Kik! Kik!* calls. Like all of our woodpeckers, they have an undulating style of flying, alternating flapping and gliding.

## How likely am I to see one?

Great Spotted Woodpeckers are fairly common, with around 140,000 breeding pairs in Britain, but are not that conspicuous. They are easiest to see in late winter and early spring at a time of year when they regularly engage in territorial drumming. Their sounds give them away - the sharp *Kik!* call - and their distinctive bounding flight draws the eye as they dash between trees, swooping upwards and sticking to upright trunks.

## What makes it special?

This is a striking species with a certain cartoon-like charm. Great Spotted Woodpeckers are a wary arboreal species and will hide from an observer on the side of a tree that is out of view, holding on with strong claws spread forwards and backwards to enable maximum grip. They use their bill to good effect, digging out wood-boring grubs from beneath tree bark and excavating nest chambers in trees, steadily chipping their way inwards then downwards. The familiar drumming sound is not related to nest tunnelling, but is a territorial 'song' beat out by the male on a resonant bough in a rapid burst. Such high impact head-banging would knock them unconscious were it not for the alignment of skull bones and soft tissue that cushions the blows.

## Added interest

- Great Spotted Woodpeckers will hack their way into bird boxes to steal the eggs and chicks of species such as tits. A metal surround can protect the entrance hole from raids.
- Woodpeckers were absent from Ireland until the booming British population of Great Spotted Woodpeckers spread across the Irish Sea and began breeding in 2007.
- Great Spotted Woodpeckers famously drum with their bills on dead or hollow branches, but have also been recorded using telegraph poles, weather vanes and drainpipes.

## Not quite a Great Spotted Woodpecker?

If your Great Spotted Woodpecker looks decidedly small and lacks the bold white shoulder patch and red under the tail, then it could be a Lesser Spotted Woodpecker - an exciting find as they are scarce and declining birds. These Sparrow-sized pied woodpeckers, numbering only a couple of thousand pairs in Britain, are found in lowland woodlands in areas of England and Wales. If, on the other hand, your woodpecker is large and green, then it is a Green Woodpecker, the third of our breeding species (*see Green Woodpecker*). Finally, if your pointy-billed tree climber has a slate grey back and a black line through the eye then it could be a Nuthatch, which strongly resembles a small woodpecker (*see Nuthatch*).

**Lesser Spotted Woodpecker**

# Green Woodpecker

Female

## Key identification features

■ The Green Woodpecker is a hefty, tough-looking and distinctive bird with an olive-green back, light underside, formidable pointed bill and a blaze of red along the top of the head.

■ It has a mask of black around the pale eyes and a short black stripe runs down from the bill like a moustache - which is reddish in males.

■ The call is a loud yapping and in flight the rump is striking yellow. Young birds look like speckled versions of the adults.

## Where could you find one?

Green Woodpeckers can be found across most of Britain (though are absent from the far north and Ireland) and are most common in the southern half of the country. Large trees in which to excavate nest holes and open grassy feeding areas are the key habitat requirements, and they can be found in deciduous or mixed woodland, parkland, forest glades, fields, orchards, meadows, large gardens, and on golf courses and heaths.

## Is it easy to identify?

The Green Woodpecker may have a robust build, a fearsome dagger bill and a pale-eyed glare, but they are cautious birds that would sooner fly away than engage in a staring match with a birdwatcher. As such, a typical view is of one heading off in bounding flight, calling loudly in alarm, before swooping up onto the side of a tree - even hiding out of sight on the opposite side. If they weren't so panicky one wouldn't see half as many, as they are fairly well camouflaged in the canopy or feeding discreetly alone on the ground. Their getaway offers an excellent opportunity to positively identify them by their size, undulating flight, red cap and, most noticeable of all, the arresting splash of bright greenish-yellow on their lower back. In addition, the call, a crazed yelping laugh, is also a giveaway.

## How likely am I to see one?

Strong climber

With over 50,000 pairs scattered across most of Britain, Green Woodpeckers are resident territorial birds that don't travel far, regularly visiting favoured feeding sites, so if you find one the chances are you may come across it again in the same area. Some are lucky enough to have Green Woodpeckers drop by their back gardens.

## What makes it special?

You might imagine the largest of our woodpeckers would require sizeable prey to keep it going. Not so. Apart from the odd caterpillar or beetle, the Green Woodpecker feeds almost entirely on the tiniest of insects: ants. Digging into ant nests with its bill, it uses its long sticky tongue to slurp them up in large quantities. As a result, this eccentric woodpecker spends much of its time on the ground. However, it is a good climber, using its short stiff tail as a prop and excavates its own nest hole in a tree. Pairs, which may stay together for life, raise four to six young on a diet of thousands upon thousands of ants. Unlike the Great and Lesser Spotted, the Green Woodpecker doesn't often 'drum' its bill against a dead bough to signal its presence, but it certainly makes itself known with its penetrating 'what on earth was that?' laughing call. Learn to recognise it and you will be rewarded with a lifetime of sightings of these remarkable birds.

## Added interest

- In order to extract tasty ants from underground nest chambers, the Green Woodpecker has a sticky tongue that is so long it runs all the way around the back of the skull when retracted.
- It is believed Green Woodpeckers learn the positions of ant nests within their territory to ensure they can get enough of them – though can face starvation in cold winters when the ground freezes.
- The Green Woodpecker was once known as a Yaffle, based on its laughing, or 'yaffling', call. The 1970s children's TV series Bagpuss included the character Professor Yaffle, a bespectacled wooden woodpecker.

## Not quite a Green Woodpecker?

In the summer, if your Green Woodpecker is a bit dull and speckled looking, then it could be a juvenile. But if your woodpecker is black and white, with red under the tail, then it is a Great Spotted Woodpecker (*see Great Spotted Woodpecker*). Finally, a screechy green bird that is becoming increasingly common in suburban areas in south-east England and beyond and might be confused at a distance because of its colour, is the Ring-necked Parakeet. This exotic Asian bird has established a wild population here in recent decades after escaping captivity. It is estimated more than 8,600 pairs now live in Britain. Instead of having the Green Woodpecker's short tail and long beak, it has a very long tail and short red beak.

Juvenile

# Nuthatch

## Key identification features

- The Nuthatch is a pointy little bird that resembles a small woodpecker.
- Roughly sparrow-sized, it has a blue-grey back, short tail and dagger-like beak. The underneath is rusty-orange, the throat white and a striking black stripe runs through the eye, as if drawn on with a thick marker pen.
- Nuthatches are active, often vocal, woodland birds that grip the sides of trees as they search for food and will also visit bird tables.

## Where could you find one?

A resolute tree-hugger, the Nuthatch is typically associated with mature deciduous species, such as oak and beech, working its way up and down searching for insects hidden in the bark. As such it is found in woodlands, parks and gardens with decent-sized trees, mainly in England and Wales and more sparsely across southern Scotland. Nuthatches will also come to garden feeders, bossing other birds and taking more than their fair share of nuts and seeds to hoard for lean times.

## Is it easy to identify?

The Nuthatch looks very much like a little woodpecker, from its pointed bill to the way it grasps vertical tree trunks, clambering along the sides of branches in jerky hops as though indifferent to gravity. Viewed from behind it can look a rather plain slate-grey bird, short-tailed at the rear and sharp-ended at the front where its body tapers into the pointed beak. But side-on this dapper woodland specialist has a touch of the out-of-the-ordinary about it that never fails to impress, with orangey-brown underparts and flanks – which are more richly coloured in males – and a conspicuous black bandit-style eye stripe. The Nuthatch is also vocal, drawing attention to itself in woodland with clear whistles and piping trills. This is no shy and retiring species, though may be hard to spot when quietly foraging in the tree canopy and is most easily observed visiting bird tables.

## How likely am I to see one?

Narrowing a nest entrance with mud

Nuthatches are widespread year-round and number roughly 220,000 pairs in Britain. They seldom travel far, a few kilometres would be long haul travel for a Nuthatch, so are slow to colonise new areas. However, their numbers are on the rise and they have relatively recently spread over the border into Scotland, though you won't find them in Ireland.

## What makes it special?

Arboreal agility, handsome looks and a no-messing character give the 'one of a kind' Nuthatch bags of appeal. It is the only bird in Britain that descends a tree trunk head-first, using strong claws to get a grip on the gnarled sides of old boughs so it can search for hidden insects from all possible angles. Rufous flanks and a dash of black through the eye add a touch of the exotic to its appearance. And its sharp shape is matched with a rather direct disposition: it can clear a bird table with a pointed stare. Few small species would risk a jab from its beak. However, although the stocky little Nuthatch resembles a woodpecker, it is not an accomplished excavator able to chisel nest holes in trees. Instead, breeding pairs occupy vacant cavities and narrow the entrances of larger holes by plastering the edges with mud in order to keep out bigger birds.

## Added interest

- The word Nuthatch derives from their early name 'nuthak' based on their habit of wedging nuts into crevices and hacking the hard casing apart with their sharp bills, making an audible tapping sound.
- Nuthatches pair for life and typically live for two or three years – though the maximum recorded in the wild is an impressive 12 years.
- Visits to garden bird tables increase during the colder months, peaking in November and December, as Nuthatches gather seeds to eat and stockpile in winter stores hidden around their territories.

## Not quite a Nuthatch?

Walking in woodland you might spot a small mottled brown bird with a white underside and fine curved bill creeping up the side of a tree in Nuthatch fashion. This is the aptly-named Treecreeper, which typically moves from tree to tree, starting near the bottom and spiralling upwards looking for invertebrates, not downwards and upside-down – the Nuthatch party trick. Widespread, with a British population of 180,000 pairs, the Treecreeper is easily overlooked, so congratulate yourself on spotting one. Another possibility is the roughly Nuthatch-sized Lesser Spotted Woodpecker, with horizontal bars of black and white on the back and a red cap – a scarce and exciting find (*see Great Spotted Woodpecker*).

Treecreeper

# Carrion Crow

## Key identification features

- The Carrion Crow – or just plain 'crow' – is a familiar, large, black bird with a grating call and a wily character.
- It is basically black all over and males and females look alike, having glossy smooth black plumage, black legs and a thick black bill.
- The crow's wings are slightly fingered at the ends and its tail unfanned looks square-ended in flight. It makes a harsh cawing sound, typically repeated three or more times: *kraaar! kraaar! kraaar!*

## Where could you find one?

Across most of Britain all year round, in habitats ranging from wild uplands, moorland and coastal areas to farmland, copses and inner cities. They mostly nest in trees, so are typically found near woodland, but will also build their nests of twigs and grass on sea cliffs and on manmade structures. In north-west Scotland and Ireland, the Carrion Crow is replaced by the closely-related Hooded Crow, which is virtually identical, save for having grey on the torso.

## Is it easy to identify?

The coal-black plumage, robust build and confident swagger are all characteristic. Though other members of the crow family share the same look, the Crow has the fewest distinguishing features. Comparative size is never straightforward unless the birds are all lined up side-by-side, ID-parade style. The Jackdaw is smaller, has a silvery-grey area around the back of the head and light eyes. The Raven (*see right*) is at the other end of the size scale with a powerful beak, a wedge-shaped tail in flight and its repeated call is a deep, resonant *kronk*. Finally, the Rook is most similar to the Crow in size and shape, but has scruffy loose feathering around the tops of the legs and a longer, more pointed beak, with a conspicuous pale grey area around the base. Rooks nest colonially in treetops and are more likely to be seen in flocks than crows.

Raven

## How likely am I to see one?

Pairs are territorial

There are over a million pairs of Carrion Crows resident in Britain, apart from in north-west Scotland where they are replaced by Hooded Crows, which have waistcoats of grey feathers. Once a pair has set up a territory they remain faithful to the area throughout their lives, which is typically four years, though the record age is nearly 20!

## What makes it special?

Carrion Crows have been widely despised and persecuted down the centuries given their reputation for stealing the eggs of gamebirds, plundering newly-sown fields of grain, their reported attacks on new-born lambs and sinister associations as scavengers of the dead. They are the perfect avian rogue: black-cloaked and hard-hearted with a bad-tempered, almost foul-mouthed, sounding call.

Crows also have a cunning streak and, if nothing else, their intelligence is worthy of admiration. An ability to solve puzzles, count, share information and use tools, has been recorded across the corvid family and such brainpower is vital to the success of these adaptable opportunists when it comes to finding their next meal. Our darkest bird is also one of our brightest.

## Added interest

- Carrion Crows will forage along tidelines and they share with the Herring Gull the habit of dropping shellfish from a height onto rocks to smash them open.
- Brainy Crows in Japan have learnt to use traffic to crush walnuts, carefully placing them on crossings when the lights turn red, then safely retrieving the opened snack the next time vehicles stop.
- Along with such species as the Cuckoo and Chiffchaff, the Crow derives its onomatopoeic name from its call, a grating cawing, having been known in Old English by the word 'crawe'.

## Not quite a Carrion Crow?

If your Carrion Crow has a pointed beak the shape of a gardening dibber, with a muddy-grey bare patch at the base, then you're looking at a Rook. These gregarious farmland birds have a slightly peaked crown which is neat and glossy, as if slicked back with a comb. Found across most of Britain, they nest communally in 'Rookeries' at the tops of tall trees and are as abundant as Carrion Crows. Ravens, on the other hand, are much bigger, with a hefty beak and diamond-shaped tail and are always great to spot. Numbering just a few thousand pairs, these impressive westerly birds of mountains, sea cliffs and moors have a spirit of wilderness about them.

Rook

# Magpie

## Key identification features

- The Magpie is a common and recognisable member of the crow family, with smart black and white plumage, a long tail and a raucous cackling call.

- It is black on the head, chest, back and tail and has a white belly and shoulder patch. In good light an iridescent blue-green sheen is visible on the dark folded wings and the tail, while in flight the outer half of the fairly short and rounded wings has a visible white patch.

## Where could you find one?

It needs trees in which to build its large stick nest and open areas for feeding, but apart from that the Magpie isn't too fussy about habitat. It can be found in wooded farmland, on moorland and along roadside verges, in coastal copses, suburban gardens and leafy city centres, though is absent from dense forests, bleak upland areas and the north-west of Scotland.

## Is it easy to identify?

The answer is black and white. With its contrasting plumage and long tail, this confident corvid is unlikely to be confused with any other species and is probably the easiest British bird to identify. Added to which the sexes look the same and don't go in for seasonal plumage changes – meaning that all Magpies look like Magpies all of the time. The tail is striking when spread in flight, tapering in a long wedge-shaped fan and some of the black feathering has an attractive glossy sheen. It has a harsh and strident call like the rattling of small arms fire.

## How likely am I to see one?

The Magpie is a widespread bird, numbering well over half a million pairs in Britain. Breeding pairs defend their territories against intruders and are typically seen together, while non-breeding Magpies gather in gangs to roost and feed, particularly during the winter when noisy flocks dozens-strong may be encountered.

**Magpies gather in noisy flocks**

## What makes it special?

One for sorrow, two for joy... 1.1 million for a conflict never to be resolved? Widely persecuted in the past, the Magpie prospered during the twentieth century and has risen in number to more than 550,000 pairs. An opportunist that will eat everything from insects and berries to roadkill and bird table scraps, it has multiplied in rural areas and conquered leafy suburbs to such an extent that it is now one of our most familiar birds. But as its population has soared, this very visible 'villain' stands accused of decimating songbird numbers by raiding nests, with calls for culling to control its population. Eggs and chicks certainly make up a small part of its diet, however studies have concluded that the Magpie's overall impact on songbird numbers is negligible compared with habitat changes that have reduced nesting and feeding opportunities for struggling species. The Magpie is a smart bird with neat black plumage, a white waistcoat and long tail. It is also an intelligent and canny species with bags of character. Monogamous pairs are very much stay-at-home couples and seldom stray beyond their home territory, typically close to where they were born, throughout their entire lives, which average three years.

## Added interest

- The Magpie gets its name from the old term 'pie' to describe a mixture, as in its pied plumage, coupled with the addition of the everyday name Margaret, shortened to 'Mag'.
- They build large stick nests high up in the fork of a tree, cemented together with mud and covered with a roof of twigs to keep their eggs safe from Carrion Crows.
- Despite their reputation, Magpies are not silver thieves. According to research they are actually afraid of unfamiliar shiny objects - however, they do hoard surplus food, burying it around their territories.

## Not quite a Magpie?

The Magpie's appearance means that it is unmistakable, however the larger Hooded Crow, which looks like it is wearing a grey body-warmer, could perhaps cause confusion. Hooded Crows replace the all-black Carrion Crow in north and west Scotland, Ireland and the Isle of Man and are also distributed across northern and eastern Europe. 'Hoodies', as they are known, were once regarded as simply a race of Carrion Crow, but are now classified as a separate species - though are so closely related that hybridisation can occur in areas where the two meet and offspring have a mix of grey feathering. Some Carrion Crows also exhibit Magpie-style patches of white, or the odd white feather - a partial loss of pigmentation known as leucism.

**Hooded Crow**

# Jackdaw

## Key identification features

- The Jackdaw is a small and gregarious member of the Crow family with distinctive silvery-grey feathering on the back of the head and pale eyes.
- It has a dark grey-black body, black wings and tail and a black cap on the front of the greyish head. The beak is relatively short and the wings less fingered compared with our other Crows.
- In flight, it beats its wings quite rapidly and emits a loud *Jack!* call.

## Where could you find one?

From towering sea cliffs and wooded farms to town parks and village gardens, the Jackdaw is at home in a range of habitats. Colonies nest in the crevices of rock faces, in hollow trees, old buildings and chimney pots and feed on the ground in open areas, such as short-cropped pasture, mowed lawns, along the seashore and they will also visit bird tables. They can be found across Britain year-round, though are absent from exposed uplands and the far north-west.

## Is it easy to identify?

Even at a distance the Jackdaw looks noticeably small for a Crow, with a less ragged outline and flappy, almost pigeon-like, wingbeats. Among shadowy congregations of larger Rooks or Carrion Crows foraging in fields, it looks neat and petite by comparison and gives itself away by shouting out its name: *Jack!*

Closer views reveal a distinctive grey area covering the back of the head and neck and light grey-blue eyes. Jackdaws are social birds, seemingly with a sense of fun – flocks appear to enjoy riding the wind, especially around cliff faces or before going to roost, swirling and tumbling together while emitting their resonant calls.

## How likely am I to see one?

Jackdaw colony

Common and widespread, the Jackdaw population in Britain is an impressive 1.4 million pairs - and rising. Numbers of these opportunists increased during the twentieth century and urban living seems to their liking. In general they prefer company and stick together in loose colonies or gather in large flocks in the autumn and winter.

## What makes it special?

Perhaps it is their light eyes and jaunty character, but we tend to be more forgiving towards Jackdaws, compared with darker and more shifty-looking Carrion Crows. Yet they can be a nuisance, particularly when colonies build their nests in chimney pots. Their tactic involves haphazardly dropping sticks down a flue until enough of them wedge, then adding more and more twigs to construct a solid base on which to build their wool-lined nest. Filling a chimney with such densely packed kindling, blocks air flows and creates a fire hazard, though one has to admire the effort they put into the task. Not only do Jackdaws go to great lengths to set up home, but also to raise their chicks. Feeding a growing brood on mainly ground-dwelling invertebrates is a full-time job for parent birds and they work together as a team. In fact, Jackdaws forge incredibly devoted partnerships. Males and females pair for life and even stick together within a flock. They perch together beside their nest, preen one another and fly side by side whenever the colony takes to the wing. DNA studies show that faithful Jackdaw pairs raise a brood in the knowledge that all their offspring are their own.

## Added interest

- Jackdaws can be spotted riding the backs of sheep in spring - pulling out wool which they use to line their nests. In return they remove ticks and parasites from livestock.
- Once simply known as 'daws', the addition of 'Jack', which conveniently describes their call, was simply a term of familiarity that also traditionally denoted this corvid's relatively small size.
- Devoted Jackdaw pairs form lifelong monogamous partnerships, remaining together through good times and bad, even if they fail to successfully raise chicks.

## Not quite a Jackdaw?

Confusion species are the Carrion Crow and the Rook - both are larger and lack the ash-grey on the back of the head (*see Carrion Crow*). Jackdaws don't make a harsh cawing sound and fly in a more hurried fashion. If you happen to be visiting the coast of Wales, the tip of Cornwall or Isle of Man and see a slender, broad-winged Crow, not much bigger than a Jackdaw, with a long curved red beak and red legs then you really have struck lucky! This is a Chough, a handsome bird of towering cliffs and grazed headlands. These social Crows with excellent aerobatic skills became extinct in England in the twentieth century, but recolonised and currently number a few hundred pairs in Britain.

Chough

# Jay

In flight

## Key identification features

■ The Jay is a stunning woodland bird, with a light pinkish-brown body and a magnificent rippled splash of bright blue on the wing.

■ Roughly pigeon-sized, it has a white area in the wings and a white rump that is very noticeable in flight, contrasting with the black tail. The head is topped with a streaky pale crown (which can be raised in a crest when alarmed) and a thick black stripe below the eye which looks like a dark shadow cast by the beak.

■ The Jay may be lovely to look at, but not to listen to as it makes a furious screeching sound.

## Where could you find one?

Widespread and common, this colourful member of the crow family is found in woodlands across Britain year-round, though is absent from uplands and northern Scotland. It inhabits mainly broadleaf woodland, especially if it contains oak, as well as coniferous forests, copses and large parks and gardens with mature trees.

## Is it easy to identify?

Despite its striking looks, the Jay is no show-off and generally keeps out of the limelight, which means that views tend to be brief and tantalising: a flash of black, white and pinkish-buff disappearing between the trees. A decent sighting is something to savour – especially that chequered bar of kingfisher-blue on the wing. The body plumage blends in surprisingly well among the colours of the treetop canopy and the Jay's distinctive feature is the one seen most often: a square white rump as it flies off. In the autumn they can be spotted collecting acorns and make flying look like hard work, paddling their broad rounded wings through the air as if struggling to swim the butterfly stroke. Jays draw attention to themselves in woodland with their hysterical screeching, especially when alarmed.

## How likely am I to see one?

Raising its crest in alarm

There are 170,000 pairs of Jays in Britain and autumn is the best time to see them. They are harder to approach in rural areas where they were persecuted for raiding the nests of game birds and may be easier to spot in urban areas – perhaps visiting a bird table for peanuts and scraps.

## What makes it special?

Exotic looking, but wary by nature, the Jay is a common bird that always feels like a rare sighting. Their ear-splitting shrieking is often directed at predators or intruders within their territory and can signal their presence in areas where you might not know they existed at all. Jays are also acorn addicts and this forces them to break cover in the autumn as they search for supplies to see them through the winter. It can seem as if a switch has been flicked: suddenly, at this time of year, Jays appear as if from nowhere. They fly to and from oak trees, collecting acorns and stashing them in scattered hiding places. By spreading their stores in numerous locations, tucked into bark crevices or buried in the ground, they reduce the risk of losing an entire hoard to squirrels or mice. Familiarity with a lifelong territory, coupled with an incredible memory, means they are able to locate their stockpiles over the months that follow.

## Added interest

- When collecting winter supplies, Jays can carry three or four acorns at a time in a throat pouch and their bill. During autumn, individuals may stash several thousand around their territory.
- While Jays generally make terrible screeching sounds, they are also capable of impressions – expertly mimicking the calls of predators such as owls, crows and even miaowing cats.
- Jays sometimes engage in 'anting', a behaviour that involves spreading their feathers over a colony of ants and stimulating the irate insects to run over them spraying formic acid, which helps rid the plumage of parasites.

## Not quite a Jay?

The Jay is a bit of a one of a kind, as no other medium-sized woodland bird has the same mix of fawn and black with the stark white rump. A female Bullfinch has the white behind and black tail, as well as a pinkish-brown body, but is much smaller and has a black cap (*see Bullfinch*). The Collared Dove has a warm buff body and dark wing tips and is a similar size to the Jay, but lacks the white rump and powerful bill (*see Collared Dove*). Finally, the Fieldfare, a large winter thrush that can be seen along woodland edges, might possibly cause confusion, but is less Crow-like and has a grey rump and chestnut brown body (*see Fieldfare*).

Fieldfare

# Farmland and wild spaces

# Yellowhammer

Female

## Key identification features

■ The Yellowhammer is an attractive finch-like bird, a little larger than a sparrow and readily identified by its bright plumage.

■ The male has a buttercup-yellow head, with a few dark streaks on the crown and behind the eye, and yellow underparts flushed reddish brown across the chest and flanks. The back is streaky brown and the dark tail relatively long and forked. In flight, a warm chestnut rump and white outer tail feathers are visible.

■ The female is plainer and more streaked with brown.

## Where could you find one?

Open spaces with scattered bushes and trees are the ideal habitat for Yellowhammers, which can be found year-round across most of Britain in heathland, farmland, young conifer plantations, coastal scrub, meadows and along roadsides, railway embankments and hedgerows. They feed predominantly on seeds, but also eat invertebrates in the summer and will occasionally visit gardens.

## Is it easy to identify?

In spring, a male Yellowhammer is at his brightest best, with canary-yellow head and handsome yellow-and-brown body, and can look more like an escaped cage bird than a resident British species. In good light this is one of our most eye-catching small birds, drawing attention not only with lemon-zest plumage but also by singing from prominent perches through the spring and summer. The rattling song, a series of notes typically ending with a drawn-out wheeze, is depicted by the well-known mnemonic *little-bit-of-bread-and-no-cheeeese* – appropriate for a bird the colour of butter. They can be wary, but even when heading off in jerky flight the yellow head is noticeable, as is the reddish-brown rump and white outer tail feathers. Males are less vivid after moulting in the autumn, while the females are duller by comparison year-round.

## How likely am I to see one?

A keen seed eater

Yellowhammers are relatively common and widespread birds, numbering over 700,000 pairs in Britain, and can be encountered in open and bushy lowland rural areas, particularly in southern and eastern Britain. They tend not to travel far, though in winter will roam in small flocks searching for seeds, often in the company of other buntings, finches and sparrows. A walk in suitable habitat stands every chance of a sighting and they can be detected in spring by their repetitive song.

## What makes it special?

A cheery ray of sunshine even on a dull day, the Yellowhammer lifts the spirits. This familiar species adds a splash of colour to country scenes up and down the land, whether perched on a hedgerow or scrabbling around for weed seeds at the edge of a field. The simple song, repeated unflaggingly by males to attract a mate and proclaim territorial rights, provides the soundtrack to sunny days. If it conjures up a wistful sense of summers past, that may be because the Yellowhammer was much more common in previous decades. The population has more than halved since the 1970s and the decline has been blamed on changing agricultural practices which have reduced the availability of invertebrates, weed seeds and stubble grain. It is still far from rare and the losses appear to be slowing, but concern about its future means the Yellowhammer has joined a number of other farmland species on the UK's Red List.

## Added interest

- While it may look like a finch, the Yellowhammer is actually a member of the bunting family – a group of seed-eating birds with characteristics which include a curved cutting edge on the lower bill. The German name for a bunting is 'ammer', from which the Yellowhammer originally took its name.
- Yellowhammers were once known as Scribble Larks because their pale eggs are marked with dark squiggly lines.
- While male Yellowhammers all sound much the same to an untrained ear, researchers have discovered they have regional dialects – their songs differing slightly from one area to another.

## Not quite a Yellowhammer?

If your 'Yellowhammer' looks on the small side and has a dark cap, then it could be a Siskin. These agile little yellow-and-black finches are found in woodlands (though increasingly visit bird feeders) and have a noticeable yellow bar across their dark wings and pale streaky flanks (*see Siskin*). If, on the other hand, the sleek yellow bird you are looking at constantly pumps its longish tail up and down, it could be a Yellow Wagtail. This summer visitor, mainly found on meadows and pastureland in England, has a yellow face and underparts, a thin bill – unlike the bunting's stubby seed-cracker – and an olive-tinged back which lacks the warm red-browns of the Yellowhammer (*see Grey Wagtail*).

Siskin

# Linnet

Female

## Key identification features

- The Linnet is a charming little finch with a particularly pleasant twittering song.
- The males have a grey head, a cinnamon-coloured back and smudge of deep pink on the chest and forehead, while the females are plainer and lack the rosy flush.
- In flight, the slightly forked tail has white patches on the sides and a white area is noticeable in the dark open wings.

## Where could you find one?

These seed eaters are widespread in scrubby areas across Britain, except for the far north-west and uplands. They can be found on rough ground, salt marshes, wasteland, heaths, moors, farmland field edges and coastal grasslands with plenty of seed-rich weeds and dense bushes (in particular gorse) which they nest beneath. Linnets are birds of the wild outdoors and have yet to become regular garden table visitors like the closely-related Goldfinches and Siskins.

## Is it easy to identify?

Linnets like to stick together. Whether breeding or feeding they are generally seen in small flocks in unkempt open places with plenty of weeds, tangles of vegetation and bushes. This gregarious nature can help identify them as, despite being fairly common, they are typically quite wary and hard to approach. If you spot a chatty close-knit group of finches in the right kind of habitat they could well be Linnets. And when they fly off, bouncing along together and twittering as they go, look out for a spray of white across the flight feathers, white sides to the dark tail and the males' light chestnut backs. They can even sound a little like they are saying their name *linn-et*. A decent view of a male in breeding plumage will reveal the attractive pink on the forehead and on either side of the breast.

## What makes it special?

Before imported novelties such as Budgerigars and Canaries became popular pets and laws protecting wild birds were introduced in the 1880s, home-grown species like Linnets and Goldfinches were trapped in their tens of thousands and kept as cage birds. Goldfinches seem to have forgotten our past cruelties by increasingly adorning bird tables as garden visitors. The Linnet however, remains a resolutely wild species and shuns backyard offerings, despite being a seed addict. They essentially eat seeds all year round, even taking advantage of oilseed rape crops and haven't much of an appetite for creepy-crawlies. This pretty, twittery little finch has undeniable charm, adding interest on a country walk. The fact that it generally keeps its distance makes identification all the more satisfying and close-up views something to savour.

### How likely am I to see one?

There are more than 400,000 pairs of Linnets in Britain, but you need to get out and about to spot one as they do not generally make their homes in urban areas or visit bird tables. In summer, they nest in loose colonies of a few pairs, while in winter they gather together in larger flocks, often mixing with other finches, sparrows and buntings and some even head south into Europe.



## Added interest

- The name Linnet derives from their liking for the seeds of the blue-flowering cultivated plant flax, known as 'linette' in French, from which linen and linseed oil are made.
- While small birds typically defend insect-rich territories against competition when raising their hungry broods, Linnets are happy to nest in groups as they roam more widely to find the seeds which form their staple diet.
- Linnet numbers have more than halved in Britain since the 1960s, with declines blamed on a shortage of weed seeds in the cultivated countryside, and the species is now on the UK's Red List (*see page 186*).

### Not quite a Linnet?

In north-west Britain, where Linnets are largely absent, the far scarcer Twite can be found. This upland and moorland species is darker and streakier than the Linnet and lacks the chestnut back, grey head and pink chest of a male. In summer, male Twites have some pink feathering on their rump and in winter their stubby bills turn yellow. Around 10,000 pairs breed in Britain, (mainly in Scotland).

If your 'Linnet' is hanging upside down, looks quite streaky, has a pale bar on the wing and a tiny black bib, it could be a Lesser Redpoll (*see Siskin*). This small finch does have pink on the forehead and sometimes its breast, but is a woodland species. An exciting bird to spot!

**Lesser Redpoll**

# Reed Bunting

Female

## Key identification features

- The Reed Bunting is a sparrow-sized bird with a reddish-brown, streaky back and wings and a lighter underside.
- In the spring and summer, the males have striking black heads and bibs, a white collar and short white streak drooping down from the beak like a little moustache.
- The females are less conspicuous, with pale streaks and dark patches on the head. Both have white sides to the tail, visible in flight.

## Where could you find one?

Reed Buntings are common across most of Britain throughout the year in a variety of habitats. As the name suggests they are mainly found living in reedbeds, as well as among rushes and rank vegetation near freshwater where there is plenty of the insect life needed to raise a brood or two of chicks during spring and summer. They have also colonised farmland, especially fields of oilseed rape. In winter, they feed mainly on seeds and will even visit bird tables in rural and suburban gardens.

## Is it easy to identify?

A male Reed Bunting in full breeding plumage is enough to stop even a non-birdwatcher in their tracks with the question: 'Why has that sparrow got a black head?' Their two-tone markings include a white neck-band which emphasises the jet black of the head and a white stripe running down from the base of the beak, like a trickle of milk flowing into the collar – unusual and eye-catching. Females lack the impact of the male's stand-out head, instead having a triangle of dark brown on the cheeks bordered by a pale streak below the bill and above the eye. One useful year-round identification pointer for both sexes is that when they fly away their tails look dark with noticeable white outer edges.

## How likely am I to see one?

Males repeat a simple song

There are about 230,000 pairs of resident Reed Buntings in Britain. They are easiest to see in the spring, especially in favoured wetland habitats, as the males are most showy at this time of year with their distinguished black-and-white head markings, often singing from a visible perch. The simple and repetitive song may also help locate them: a short staccato series of notes which has been variously described as sounding like: *'one, two, testing'*, *'here comes the bride'*, *'three blind mice'* and so on. Take your pick - or make up your own translation.

## What makes it special?

The Reed Bunting can make an expert of even the birdwatching beginner. To see one and know what it is, feels like a small triumph. It is not exactly a widely-known species; it looks quite a bit like a Sparrow (they were once called Reed Sparrows or Bog Sparrows) and its name has an ornithological ring to it (that word 'bunting' instead of, say, 'finch'). Yet it is a widespread bird and breeding males with their black-and-white heads are relatively easy to identify. It is also good to know they are still around in large numbers. In the latter half of the twentieth century, the UK's population of Reed Buntings plummeted at an alarming rate, with agricultural intensification thought to have been a factor. Fortunately, these adaptable birds have since recovered sufficiently to be taken off the UK's Red List and can be encountered across most of the country.

## Added interest

- They moult after the breeding season to replace worn out feathers and the male's fresh head plumage is tipped with brown, giving him a lacklustre look over winter. The ends wear away to reveal the black beneath in time for spring.
- Studies have revealed that the female Reed Bunting gets up to mischief when her partner's back is turned, with around half of her chicks fathered by other males.
- Reed Buntings rely on their acting skills to distract potential predators approaching their nests built low to the ground, by hopping away with wings half spread, pretending to be injured.

## Not quite a Reed Bunting?

The male Reed Bunting is not the only small bird with a black head, but few of the alternatives have his streaky-brown sparrow-like back. The Stonechat has a black head and white collar, though is smaller and has rusty-red on the breast. This attractive species is found in scrubby open spaces with scattered bushes and males typically perch in full view (*see Stonechat*). Reed Buntings are most easily confused with Sparrows, which have different head markings and lack streaking on the underside and white outer tail feathers (*see House Sparrow*). Finally, if your female 'Reed Bunting' looks a little on the yellow side then it could be a female Yellowhammer (*see Yellowhammer*).

Stonechat

# Stonechat

Female

## Key identification features

- The Stonechat is a plump little bird, similar in shape and perky character to a robin.
- The male has a jet black head, with a half-collar of white on the side of the neck. His back and short tail are dark mottled brown, while his front is an attractive rusty-peach.
- The female has a more muted appearance which lacks the strong contrasting colours, though has a hint of a dull white half-collar and orange on the chest.
- Stonechats perch conspicuously on the tops of bushes and fences, often flicking their tail up and down in a fidgety manner, and make a *chat-chat* sound.

## Where could you find one?

Scrubby open spaces with low scattered bushes provide ideal habitat for Stonechats, which like to perch a few feet up and scan the ground for insect prey. As such they can be found on heaths, moors, cleared forestry plantations and coastal grasslands which have plenty of heather, bramble or gorse. They are most common year-round in the west and south of Britain and can be found in eastern areas during autumn and winter.

## Is it easy to identify?

The Stonechat makes life easy for the birdwatcher by perching confidently in full view on the top of a gorse bush or fence post, saying its name: *chat-chat*. The sound is a bit like two stones being clacked together, hence 'stone-chat'. During the spring and summer breeding season, the male stands out as a handsome little bird in scruffy settings with his marmalade chest, striking black head and wedges of vicar's-collar white on either side of the neck. After moulting in the autumn, the colours are obscured by the buff tips of fresh feathers, which gradually wear away to reveal the vivid plumage beneath. In flight between lookout perches, a central light patch is visible on the short whirring wings.

## How likely am I to see one?

Stonechats typically perch in full view

The Stonechat is fairly common in suitable habitat and more than 55,000 pairs breed in Britain, mainly in the north, west and south. It can be regularly encountered on walks through bushy moors and heaths or along coast paths. In the winter, many exchange inland territories for milder coastal scrubland.

## What makes it special?

The charming Stonechat is always a satisfying bird to identify on a country walk. Putting a name to this relatively obscure-sounding species may sound impressive, but in truth it does its level best to be recognised. Appearing on a prickly gorse bush, it poses the quick-fire question: 'What am I?', then hangs around offering up clues, restlessly twitching its wings and tail as if impatient for an answer. Repeating its name *chat-chat*, it holds its position with the clock ticking while you get your binoculars in focus, the male showing off his black head, white mark on the neck and orange-red breast. Time's up! Off it flits, but seldom further than the next convenient perch, drawing attention to itself with its hard calls in case you have trouble relocating it. The Stonechat is truly a friend of the beginner birdwatcher and expert alike.

## Added interest

- This small insect-eater suffers significant losses during hard winters, but populations are able to bounce back in good years as pairs can raise three broods of five eggs.
- Male Stonechats stand guard on lookout posts and sound the alarm if they spot predators, acting as useful sentries for other birds which may live close by in order to take advantage of their vigilance.
- Some of our Stonechats fly south to Mediterranean areas in the winter rather than tough it out here through the colder months.

## Not quite a Stonechat?

If your 'Stonechat' has a white eyebrow then it is likely to be a slightly scarcer Whinchat, which lives in similar kinds of rough and uncultivated places and is also a good bird to spot. More than 45,000 pairs of these summer visitors breed in northern and western upland areas on slopes of bracken, bushy grasslands and moorland fringes. Like the Stonechat, closely-related Whinchats will perch in full view and the brighter male has eye-catching white borders around his black cheeks and an apricot chest. In flight, white sides are visible at the base of the tail. Loss of habitat means these long-distance migrants, that winter in Africa, have suffered worrying declines and are on the UK's Red List (*see page 186*).

Whinchat

# Wheatear

Female

## Key identification features

■ The Wheatear is a small and attractive species of rocky and grassy open areas.

■ The male is a handsome bird, with a neat grey back, pale underparts, black wings, a dark patch around each eye topped with a white eyebrow and a flush of light apricot on the breast. The female is more toned-down and sandy brown in appearance.

■ Both sexes share a stand-out feature: a white tail marked at the tip with a broad black upside-down 'T' shape.

## Where could you find one?

This ground-dwelling summer visitor favours areas with short-cropped turf, where it can find invertebrate food, coupled with boulders or drystone walls where it can hide its nest. As such it is found on downland, in grassy coastal areas and especially across northern and western uplands grazed by sheep and rabbits – and will even nest in disused burrows.

## Is it easy to identify?

If you have never seen a Wheatear before then the male's striking looks when he is at his brightest in spring may come as a surprise. Foraging on a coast path or rocky upland slope, with his orange-flushed front, smart blue-grey back and black face mask, he can appear like some windblown rarity from overseas that simply couldn't belong in Britain. The Wheatear's tell-tale tail is a key to identification: a distinctive flash of black and white, seen as it flies away, generally low to the ground and not very far. The feature, shared by both sexes, has a geometrical quality, like two interlocking pieces of a puzzle and is impossible to ignore. Given they are fairly tolerant of people, one can often enjoy decent views of these upright birds as they hop and run over the turf searching for insects.

## How likely am I to see one?

Nests among rocks

If you go walking across uplands in the north and west of Britain during the spring and summer you are pretty likely to see a Wheatear - particularly around drystone walls where they typically nest. Some 230,000 pairs breed here and they also turn up around the coast and in lowland areas on migration.

## What makes it special?

The Wheatear is a species that people notice - even those with no interest in birds. The reason is its rear end, which has evolved to startle and confuse predators: in flight, it fans out to reveal an eye-catching black-and-white pattern of road sign simplicity. This is a bird of uplands, moorlands and coastal grasslands and to know its name, an unfamiliar-sounding one at that, can make you feel like a proper birdwatcher. 'Ah, that'll be a Wheatear', you can declare to puzzled companions and, however cold and wet it is, you can bask in a warm glow of pride for the rest of the walk. Our visitors spend the winter in tropical Africa and come here to breed during the spring and summer. However, some travel as far north as Greenland, flying non-stop thousands of miles over the Atlantic to southern Europe on their return journey - a huge test of stamina for such a small bird.

## Added interest

- The Wheatear is one of the first spring migrants to arrive back in Britain, getting here in March ahead of swallows, swifts and other summer regulars.
- The name Wheatear derives from the Old English term 'white arse', which describes its conspicuous tail plumage.
- Wheatears used to be trapped in the autumn in southern counties and eaten in large numbers, and were considered a delicacy in Victorian times, despite their diminutive size.

## Not quite a Wheatear?

On moorland, heaths or rough grassland you might be confused by a smaller bird: the Whinchat. The dark eye patches, white eyebrow and orangey breast seem similar, but the Whinchat has a streaky brown back and the black and white on its tail is not so well defined. It tends to perch on bushes and small trees, unlike the ground-loving Wheatear (*see Stonechat*). And if you come across what looks like a male Wheatear climbing up a tree trunk, that will be a Nuthatch - a woodland dweller and garden visitor which shares the grey back and some rusty-coloured feathering underneath, but has a long black stripe through the eye, lacks white in the tail and behaves more like a woodpecker (*see Nuthatch*).

Whinchat

# Meadow pipit

## Key identification features

- The Meadow Pipit is a small, slender bird, streaky brown on top and pale beneath with dark dashes on its chest and flanks.
- It has a thin insect-catching bill, pinkish legs and white outer tail feathers which may be visible in flight.
- An inconspicuous ground feeder, the Meadow Pipit rises weakly into the air when disturbed and makes squeaky peeping sounds.

## Where could you find one?

The Meadow Pipit is an abundant bird of open spaces across Britain, from upland moors, heaths and rough grasslands to salt marshes, dunes and coastal meadows. Most common in the north and west during the spring and summer breeding season, this resident species abandons high ground in the winter and can be found more widely in lowland areas, on farmland and near the coast.

## Is it easy to identify?

This nondescript streaky brown bird might pose the ultimate identification challenge – were it not for the fact that it is so common, especially in moorland and upland areas. Meadow Pipits feed discreetly on the ground, walking in a jerky fashion and when flushed they rise up in feeble flight, seemingly struggling to gain height before dropping back down again, all the while making thin peeping sounds that can hint at their 'pipit' name. During the breeding season, Meadow Pipits perform territorial flight displays – climbing steadily, uttering a series of accelerating notes, before descending in song with wings spread as if parachuting back down to earth. Their plumage incorporates subtle olive-brown and golden hues, while their front and flanks are punctuated with dark hyphens.

## How likely am I to see one?

Perching beside open farmland

Meadow Pipits are common throughout the year, with two million pairs breeding in Britain. In upland areas during the summer they can be the most abundant birds around and a constant companion of the hill walker. In the autumn numbers are boosted by young birds and flocks of migrants visiting from Scandinavia.

## What makes it special?

These streaky, squeaky little birds are not exactly the pick of the bunch when it comes to looks and singing ability. Their lack of colour and standout features count against them in the beauty stakes and they would struggle to get past the first round in an avian version of X Factor. They are also fairly low down the natural pecking order, being a staple prey item for a whole host of raptors, including Merlin and Hen Harrier, while their nests on the ground are relentlessly plundered by foxes and stoats or parasitised by Cuckoos. Yet this plain, put-upon pipit, is anything but a Darwinian dead-end: it is an extremely successful species which can be found across the country year-round and numbers in the millions. Delicate and neurotic they may appear, but Meadow Pipits are tough enough to nest in some of our remotest upland areas and survive through our freezing winters.

## Added interest

- The Meadow Pipit has a very long curved claw on its hind toe that is thought to provide stability when standing on soft ground.
- It is one of the species most often targeted by Cuckoos, which sneakily lay their eggs in nests and leave duped pipit pairs to raise the outsized young as if they were their own.
- During spring and summer Meadow Pipits are abundant in our uplands, but come the winter they desert high ground and migrate downhill to milder lowland areas and coasts in search of food.

## Not quite a Meadow Pipit?

Similar pipits are the Tree Pipit and the Rock Pipit. Over 80,000 Tree Pipit pairs arrive from Africa in April and depart by September and have finer and thinner streaks on the sides of their body. The Rock Pipit, numbering 35,000 pairs, is found year-round along rocky shorelines. It is muddier brown with mottled smudgy streaking on its underparts and has dark legs.

If your 'Meadow Pipit' is large, has a crest and a white trailing edge to the wings, it could be a Skylark (*see Skylark*). Finally, if it's grey and plain, it might be a Spotted Flycatcher. A widespread summer visitor with subtle streaks on the head and pale grey breast, it numbers around 33,000 pairs. A great bird to watch in fly-catching action.

Spotted Flycatcher

# Skylark

Crest raised

## Key identification features

■ The Skylark is a streaky sandy-brown bird, shaped a bit like a small thrush.

■ Its back is dull brown and mottled with dark streaking, while its underside is paler and finely streaked on the golden-flushed breast. It has a short crest that it sometimes raises and pale pinkish legs.

■ In flight, white outer tail feathers are visible and the trailing edge of the wing is also lighter. Males climb steeply in song flight during the breeding season, singing continuously as they ascend, hover and descend.

## Where could you find one?

A common resident bird of open countryside, the Skylark is widespread in all kinds of exposed places, from arable fields and grasslands to heaths and moors. It nests on the ground, hidden within short vegetation away from hedges or woods and males sing in the sky over their territories. In the winter, upland birds join lowland flocks feeding on winter stubbles, in grassy fields and around saltmarshes and at this time of year they are most common in the southern half of Britain.

## Is it easy to identify?

The Skylark is a well camouflaged ground-dweller that seldom perches in view – perhaps occasionally on a fence post – and can be hard to spot feeding amid moorland, farmland or meadowland grasses. When flushed, it flies off low with quivering wingbeats uttering a short purring chirp. Pale tracing at the rear of the triangular wings can be seen, along with white sides to the tail. Both sexes have a blunt crest of feathers on the head, which certainly helps with identification, but this is often not raised. Given its fairly nondescript appearance, this acclaimed singer is best identified in song flight, when the male climbs in full voice on fluttering wings, holding his position high in the sky before returning to earth.

## How likely am I to see one?

A species of open grassland

This iconic farmland and grassland species can be found the length and breadth of Britain and in spring and summer their songs fill the air over open ground. A country walk on a sunny day is sure to offer up a Skylark or two - though a silhouetted singing bird at altitude can be tricky to locate in the sky. Some 1.4 million pairs nest in Britain and in the autumn and winter numbers are swelled by visiting birds from northern Europe.

## What makes it special?

A Skylark may not be much to look at, but is certainly a lot to listen to as it rises from the ground in uplifting song, showering its surroundings with an unbroken stream of rapid trills and warbles like an overwound music box. Its song has seduced poets and musicians down the centuries and it was caught in vast numbers for the Victorian cage-bird trade - as well as the cooking pot. Trapping Skylarks has long been outlawed in Britain, but life has been far from easy for this much-loved species over recent decades. Changing agricultural practices, including the autumn sowing of grain crops and regular cutting of grass for silage, have deprived Skylarks of suitable nesting habitat and winter food, so numbers have fallen steeply. Enjoy the Skylarks you do hear - they're what skies were made for.

## Added interest

- When not singing, Skylarks are reluctant to take to the air and on being approached will initially crouch low to the ground rather than fly.
- The exhausting act of singing in flight advertises a male Skylark's fitness to a potential mate, helps ward off rivals and also displays his courage - given that aerial performances also draw the attention of predatory falcons such as Merlins.
- A male Skylark generally sustains his song flight for anything between a couple of minutes and a quarter of an hour, though renditions more than twice this long have been recorded.

## Not quite a Skylark?

On the small side, lacking any hint of a crest and making peeping sounds when flushed from the ground? It could be a Meadow Pipit (*see Meadow Pipit*). If your Skylark is singing perched in a tree then it is certainly no skylark - though could perhaps be a Tree Pipit or a scarce Woodlark, which has a white eyebrow, short tail and a melancholic fluty song typically given in circling flight. Finally, if your Skylark has a plump body, stubby bill and is delivering its jangling song from a farmland hedge-top or fence post, then it might be a Corn Bunting. Found in eastern and central lowland areas and numbering around 11,000 pairs, it is an excellent species to spot.

Corn Bunting

# Fieldfare

## Key identification features

■ The Fieldfare is a large, stocky and handsome thrush, with a grey head and rump, black tail and a chestnut back.

■ Its light underside is densely spotted on the chest, which has a rusty-yellow tinge, and along the flanks.

■ In flight, the underwings are white and the grey lower back stands out, sandwiched between the dark tail and brown back. It is a gregarious bird and flocks communicate with hard calls: *chack-chack-chack*.

## Where could you find one?

A widespread winter visitor, arriving from October onwards and departing by spring, the Fieldfare can be found throughout Britain in open fields, along hedgerows and woodland edges, on farmland, parkland and playing fields. In hard winters they may turn up in large gardens to plunder berry bushes and feed on windfall fruit.

## Is it easy to identify?

This sociable migrant is almost always found in flocks and will mix with other thrushes feeding on worms and other invertebrates in fields or on hedgerow berries. The Fieldfare can be readily told apart by its blue-grey head, pale grey rump and reddish-brown back, as well as its cackling chuckle. It is a vocal species and this call draws attention to ragged flocks flying overhead and helps identify distant gatherings silhouetted on bare branches or obscured by winter mists. Close up it is an attractive deep-chested, chestnut-topped bird, its head a distinguished pewter grey, and breast stained the colour of dark honey and dabbed with dark spots.

## How likely am I to see one?

Plunders winter berries

Around 700,000 Fieldfares spend the winter in Britain before returning to their breeding grounds in northern and eastern Europe. A walk on a frosty day in open countryside stands every chance of an encounter and occasionally they can be tempted into gardens, particularly during snow, with offerings of cut up fruit.

## What makes it special?

The appearance of Fieldfares in late autumn heralds the coming of winter and yet these harbingers of hard times seem upbeat birds. Robust and upright, they march in loose formation across damp fields, hopping in a stop-start fashion in search of worms, beetles and other invertebrates, or descend on berry bushes to feast on the banquet. Chattering as they go, drifting flocks of Fieldfares lift the spirits on wet and chilly days – they always feel like special birds, despite being common winter migrants.

Shortages of autumn food push the first arrivals across the North Sea in October. They are accompanied by smaller Redwings, close relatives that have a patch of red on the flanks and white eyebrows. Often seen together, these wanderers comb our countryside for food until the days begin to lengthen, the Arctic snows thaw and spring calls them back to their homelands to breed. A Fieldfare in the summer is a very rare sight indeed – let a local expert know about it!

## Added interest

- The name Fieldfare dates back to Anglo-Saxon times, when this nomadic winter visitor was called the 'feldefare', or 'traveller over the fields'.
- Fieldfares breed in Scandinavia, central and northern Europe. However, two or three pairs are found nesting in northern Britain every year – though the species has yet to put down permanent roots.
- On regular breeding grounds they defend their nests by noisily dive-bombing potential predators and showering them with well-aimed excrement.

## Not quite a Fieldfare?

If the bird you have spotted shares the upright stance, heavily spotted front and white underwing of the Fieldfare, but is missing the chestnut back and light grey rump, then it could be a Mistle Thrush. Like the Fieldfare, this similar-sized species will also feed far from cover in the centre of fields or gorge on berry bushes, but doesn't gather in large feeding flocks (see *Song Thrush*). If, on the other hand, the streaky-breasted thrush in view has rusty-red sides and – most noticeable of all – a distinctive white eyebrow, then it is a Redwing. Around 650,000 Redwings spend the colder months in Britain and flocks forage in open grassy areas keeping in touch in flight with thin high-pitched calls.

Redwing

# Cuckoo

Female

## Key identification features

■ The Cuckoo is a dove-sized species famed for the song from which it takes its name and for laying eggs in other birds' nests.

■ It looks vaguely like a cross between a pigeon and a hawk, with grey plumage and pointed wings. Blue-grey on the back and chest, its long dark tail is spotted with white. The pale underparts are finely streaked with black barring and the eyes and feet are yellow. It flies with rapid, shallow wingbeats, typically holding its pointed wings below the horizontal.

■ Females, which make a bubbling trill sound, have a flush of rusty-brown on the chest, while juveniles are browner all over and have a small white patch on the back of the head.

## Where could you find one?

The Cuckoo can be found across Britain in the spring and summer in all kinds of habitat, from moorlands and marshes to open woodland, farmland and parks. It is one of our most widespread summer migrants and draws attention to itself by repeatedly calling out its name.

## Is it easy to identify?

The Cuckoo's loud and almost comical *cuck-oo* is unmistakable when heard, is repeated for good measure and is only delivered during the breeding season when these summer migrants are present in our country. They have short legs, so appear to squat on a perch and often droop their wings and cock their round-ended tail up slightly. First impressions are of a pigeon-like bird that is grey on top and has a white belly marked with horizontal barcode lines. In flight, it can resemble a small bird of prey such as a Kestrel, given its pointed wings, small head and long tail – and is often mobbed by little birds much in the same way as a raptor. However, the Cuckoo flickers its wings below the body as it flies and points the way with a bill that is not hooked like that of a falcon.

## How likely am I to see one?

Juvenile Cuckoo

Cuckoos are more often heard than seen, though a downturn in numbers means that these declining vocal harbingers of spring have sadly fallen silent in many areas over recent decades. An estimated 15,000 pairs breed across Britain. Adults are present from April to August, while juveniles depart by September.

## What makes it special?

The Cuckoo is well known for its dump-and-run attitude to parenting, and how it gets away with this underhand behaviour has fascinated naturalists down the centuries. Cuckoos arrive in Britain in April, scattering far and wide. After mating the female turns birdwatcher, looking out for the nest of a suitable host species, such as the Dunnock, Reed Warbler or Meadow Pipit. This is far from random, as females inherit a preference for certain target species and lay eggs that closely match the shell colour and patterning of their chosen victims. When the coast is clear she sneaks in, rapidly lays a single egg and removes one of the host's own clutch. The fast-growing Cuckoo embryo typically hatches first and the tiny nestling then heaves all the other eggs over the side of the nest to eliminate any competition for food. With plaintive squeals and a huge open gape, it entices the parent birds to feed it, rapidly gaining weight until it dwarfs them in size and outgrows the nest. Eventually it goes its own way (foster parents presumably none the wiser) and migrates south to join adult Cuckoos that have already made it to tropical Africa for the winter.

## Added interest

- A female Cuckoo will lay a single egg in up to 25 different nests during the breeding season – diligently removing the egg which she replaces with her own close imitation, in order to outwit those keen-eyed hosts that can count.
- With a resemblance to falcons, Cuckoos are frequently harassed by mobs of anxious songbirds seeking to drive them out of their territories, which can help identify the nest sites of potential hosts.
- Cuckoos eat the kinds of hairy, foul-tasting and brightly-coloured caterpillars that many other birds avoid.

## Not quite a Cuckoo?

The Cuckoo could be confused with a grey pigeon of some kind, though is slimmer with a longer tail. It also superficially resembles a small bird of prey, such as a Sparrowhawk, which has a similar grey back (*see Sparrowhawk*), or a Kestrel, which has pointed wings and a longish tail. A Kestrel is also much the same rufous colour as a juvenile Cuckoo or a particularly brown female Cuckoo (*see Kestrel*).

Sparrowhawk

# Pheasant

Female

## Key identification features

■ The Pheasant is an outrageously extravagant gamebird with an extremely long tail and glossy body feathers the hues of polished copper, bronze and gold, infused with plum purple and hints of grey-green. Some feathers are tipped black or marked with white.

■ The head and neck are a dark metallic bluey-green and the face has red wattle patches around the eyes. Short ear tufts may be visible at close quarters and many have a thin white collar that rules a line between the head and body plumage.

■ Females are light brown, mottled with darker markings and also have a long barred tail. Juveniles look like females, though the tail is shorter.

## Where could you find one?

The Pheasant is common and widespread across Britain all year round, though absent in the far reaches of north and west Scotland. It is a ground-feeding bird of lowland farmland and woodland edges and can be found in areas with hedgerows, reed cover, thickets and copses – and dicing with death on roadside verges!

## Is it easy to identify?

An adult cock Pheasant is a large and richly patterned bird, sporting a preposterously long tail and a head adorned with ear tufts and face wattles, such that it would be hard to mistake for anything else. Even in silhouette the heavy body, stiff tapering tail, thick neck and small head are instantly recognisable. They are also vocal birds, especially in the spring, uttering resonant guttural sounds generally followed by an audible burst of rapid wing flapping. They can be flushed from underfoot, rising noisily and almost vertically, though lift-off for such a heavy bird is hard work and their short rounded wings, which alternate rapid whirring with glides, seldom take them far. Females may seem plain in comparison to males, but close up their camouflage patterning is subtle and attractive and they share the diagnostic long tail.

## What makes it special?

For such an incongruous bird, the non-native Pheasant, of Asian origin, has become a familiar feature of our countryside. Males are undeniably stunning and a decent view reveals exquisite detail in their lustrous plumage. A tail feather chanced across on a country walk simply has to be picked up and brought home. However, the Pheasant is generally viewed with derision as a bird big on looks and small on brains, given its baffled expression, suicidal road sense and readiness to fly headlong into volleys of gunfire. Although wild Pheasants roam freely, huge numbers are reared and released every year by shooting interests and the historic protection of copses and wooded habitat on farmland for this bird has benefited a wealth of other wildlife. The eccentric-looking Pheasant may not appear entirely at home in our fields and woods, but they have been here for a couple of thousand years. Initially imported as table birds by the Romans and later the Normans, feral populations were well established by the fifteenth century. Numbers later increased with the growth of the shooting industry in the Victorian era and in recent decades captive-bred releases have risen in response to the decline of native Grey Partridge quarry.

### How likely am I to see one?

There are perhaps 3-4 million Pheasants living in the wild in Britain, which is boosted by the release of some 35 million reared birds onto estates in advance of the four-month-long winter shooting season. Lacking survival skills, a quarter will die before the season even opens on 1 October, with many taken by foxes or killed on the roads.

## Added interest

- Male Pheasants can attract a small harem of females, though only high-testosterone territorial individuals are successful. Those with the largest black-speckled red wattles, ear tufts, tails and leg spurs are believed to be the most attractive to females.
- Female Pheasants lay enormous clutches of eggs, numbering between 10-14 and rely on the camouflage of their mottled buff-brown plumage to avoid detection as they sit tight on the ground nest.
- Pheasants spend their days on the ground - even preferring to run rather than fly from danger - but roost at night in sheltered trees, out of reach of predators.

### Not quite a Pheasant?

The plumage of Pheasants can vary, given the introduction of differing races and hybridisation between breeds. Light-coloured birds, dark purple-black individuals and even white Pheasants can be spotted roaming the countryside. Juvenile Pheasants look like females, but have shorter tails and may be confused with smaller partridges (*see Grey Partridge*).

Finally, on upland heather moors in the North and West, the Red Grouse could cause confusion. This plump gamebird, with a population of 230,000 pairs in Britain, has mottled chestnut brown plumage and flies rapidly on rounded wings, though unlike the Pheasant its tail is short and dark. Red Grouse have feathery feet and the males sport striking red eyebrow wattles.

**Red Grouse**

# Grey Partridge

Female

## Key identification features

■ The Grey Partridge is a ground-dwelling gamebird, similar in size to a pigeon but plumper and with broad, short wings.

■ From a distance it looks rather plain, however good views reveal a range of attractive markings. The face is orange, the neck and breast grey, the back streaky light brown and the flanks are hatched with broken bars of chestnut brown. There is a dark brushstroke of creosote brown on the belly which is more distinct on the slightly larger males.

■ Flight is typically over a short distance, alternating rapid flaps and glides and reveals rusty orange sides to the tail.

## Where could you find one?

Well, not in a pear tree for starters. This highly terrestrial bird seldom leaves the ground, preferring to get about on foot rather than stretch its wings. Grey Partridges live in grassy fields, meadows, and on arable land bordered by scruffy vegetation and hedgerows under which they nest. These year-round residents are widely, but sparsely, distributed across lowland areas and mostly concentrated in central and eastern England, as well as eastern Scotland.

## Is it easy to identify?

This shy farmland bird is easy to overlook as it feeds in and around fields, hunched close to the ground among low vegetation. Dull plumage colours blend in with its surroundings and flocks only take flight as a last resort, preferring to keep a low profile and run for cover if threatened. The Grey Partridge is much more thinly spread than in the past, so every sighting of this wary species is something to celebrate.

One is most likely to come across them by accident, or they may catch the eye when driving through suitable arable habitat: rotund grey-breasted birds discreetly foraging in the weedy stubble. The orange face is a key feature, along with a croissant-shaped patch of dark brown on the belly. To confuse matters there is another species of partridge in Britain: the introduced Red-legged Partridge (*see right*).

## How likely am I to see one?

We have lost more than 90 per cent of our Grey Partridges since the 1970s and the current population is estimated at 43,000 pairs. These nervy birds are most easily spotted at dawn and dusk, when their creaking calls can give them away. Good views may also be had by scanning suitable fields from a parked car.

Calling at dawn

## What makes it special?

A much-cherished native gamebird, the Grey Partridge was once one of Britain's most abundant species, with shooting bags during the late Victorian era and early twentieth century reaching a staggering two million birds a year. Arable fields bordered by scruffy vegetation and hedgerows provided ideal habitat, while the control of predators on estates helped this ground-nesting species to prosper. However, the post-war intensification of agriculture caused our Grey Partridge population to plummet.

Modern farming methods reduced hedgerow cover, insect life and seed-rich weeds and stubbles. The result is that females struggle to find suitable places to nest and their broods of growing chicks are starved of the invertebrate food upon which they depend. Despite ongoing efforts to turn around the fortunes of the Grey Partridge, numbers continue to decline and it is on the UK's Red List of birds of conservation concern. Consider yourself lucky to spot one – this iconic farmland species can no longer be taken for granted.

## Added interest

- Grey Partridges lay the largest clutches of any of our birds, averaging between 13-16 eggs, up to a maximum of 20.
- One of our least adventurous species, the Grey Partridge lives its entire life in a territory that can cover just a couple of fields and only ever roam a few hundred metres from its birthplace.
- Families stick together during the summer and autumn in groups called 'coveys' and birds roosting in the open may huddle alongside one another facing outwards to watch for predators.

## Not quite a Grey Partridge?

If your Grey Partridge lacks an orange face and has, instead, a white throat bordered by a black line which runs through the eye, then it is a Red-legged Partridge. This attractive gamebird has white-and-brown barring on the sides as well as red legs and a red bill. Successfully introduced from Europe in the 1770s, it is also known as the 'French partridge', while our native Grey is referred to as the 'common' or 'English' partridge. Red-legged Partridges are reared and released in vast numbers every year by shooting estates and have spread widely in dry lowland areas and become well established, with around 82,000 pairs now breeding in the wild.

Red-legged Partridge

# Birds of open skies

# Swallow

In flight

## Key identification features

- The Swallow is a slender, streamlined, aerial insect-feeder with a noticeably forked tail. It has a dark back that has an iridescent deep blue sheen and is off-white underneath.
- Rusty red around the face and throat is visible if you get a close view. Graceful and agile in flight, the Swallow swoops low and twists and turns as it chases insects and has pointed swept-back wings.
- The outer tail feathers of males have developed into long thin streamers, which are shorter in females and virtually non-existent in juveniles.

## Where could you find one?

Perched on a telephone wire or hunting insects over meadows, pasture and open water, the Swallow is found wherever there is a plentiful supply of flies and suitable structures where it can build its mud nest on ledges under cover, such as barns, bridges and outbuildings – farmland is ideal, with livestock fields serving up an abundance of in-flight invertebrate snacks. Swallows are also a familiar sight in villages and over sports pitches, flying to and fro with boundless energy looking for food.

## Is it easy to identify?

Swallows may be confused with those other summer sky-riding insect catchers, the martins and Swifts, but key to identifying Swallows are the tail streamers. They may not be obvious in level flight, however as the bird veers its tail briefly spreads to show pencil-thin elongated outermost feathers. They give the sense of high-velocity prowess, as if its tail has been stretched by sheer speed – although Swallows are not actually as fast as their tapered shape might suggest. From below, an attractive fan of white dots can often be spotted under the tail. Unlike House Martins, which have white rumps, the Swallow is dark across all of its upperparts. Swallows also hunt at a lower level and drink in flight from the surface of lakes and rivers.

## How likely am I to see one?

Swallows gather to migrate

Swallows are widespread in Britain during the warmer months. The first Swallow of spring, most likely to be an impatient male, is the one we tend to notice, but as their numbers build, reaching more than 750,000 pairs in Britain, they become such a familiar sight we can begin to take them for granted. That is until autumn comes and all of a sudden they've gone and the chattering sky falls silent.

## What makes it special?

Summertime and the livin' is easy – a time of year when our days are long and the skies are full of insects. That is why this annual visitor makes the perilous journey thousands of miles from wintering grounds in southern Africa to breed here. The arrival of these cheerful birds lifts the spirits: the earth is still turning, nature is in harmony with the seasons and good times lie ahead. Their migration is a remarkable feat of endurance given their small size, and their extraordinary ability to navigate back to their birthplace still confounds us. But mortality is high. More than half of all young Swallows that set out from Britain will fail to survive their African odyssey, succumbing to exhaustion, stormy weather, predation and the hazards of crossing the Sahara. Every Swallow that returns is a small cause for celebration.

## Added interest

- It was once thought that Swallows hibernated in winter, but it eventually became apparent that they flew south. It was not clear where until a British ringed bird was caught in South Africa in 1912.
- While Swallows couple up as pairs, they can be promiscuous and male Swallows with the longest, most symmetrical outer tail feathers are by far the most attractive to females.
- Swallows have a number of specific warning calls. Males worried about being cuckolded may even falsely sound the alarm to ensure their mates remain in view.

## Not quite a Swallow?

If your Swallow lacks long outer tail feather streamers it may still be a Swallow – only a young one. Alternatively, it may be a similar-looking House Martin or Swift. The House Martin has a white belly and throat and a very noticeable white rump, while the Swift has long sickle-shaped wings, looks dark all over and can be seen screaming over rooftops in urban areas during the summer (*see House Martin and Swift*).

House Martin

# House Martin

In flight

## Key identification features

■ The House Martin is a summer visitor which builds cup-shaped mud nests under the eaves of buildings and hunts insects on the wing.

■ It is similar in appearance to the Swallow, though smaller than its close relative. Its underside is pure white, the upperparts a glossy blue-black and it has a conspicuous square white rump.

■ The tail is forked, though lacks the long thin outer feather streamers of the Swallow and its pointed wings are short by comparison.

## Where could you find one?

This aerial insect-eater can be found across Britain from spring to autumn, though is scarce in the far north and west of Scotland. It nests under bridges and the eaves of buildings in villages, farmsteads and towns and needs a ready supply of wet mud close by in order to build and repair its cup-shaped constructions, as well as ample insect life to raise its chicks.

## Is it easy to identify?

Agile and streamlined species that comb our summer skies for airborne insects can seem tricky to tell apart. But the House Martin makes things easy – it has a square of white on its rump. Just a glimpse of this badge on the lower back is enough to identify it – especially as it swoops low over water, or visits a nest site. Looking up at an airborne House Martin, the dark-backed body is pure white underneath, while the similar Sand Martin has a brown breast band. The forked tail and wings are less elongated than those of a Swallow and it also tends to hunt insects at a greater height, with a more fluttering flight interspersed with glides. Its wings are also far shorter than the thin scythes of the dark-bodied Swift. In addition, the short cheery call, *prrrit,* can help identify a flock of House Martins overhead.

## How likely am I to see one?

Mud nest under eaves

Look up on a summer's day and you're likely to spot a House Martin as there are over half a million pairs breeding in Britain. Small colonies nest on vertical outside walls, under building overhangs and occasionally on natural cliff faces. Pairs can be persuaded to take up residence by fitting artificial cup-shaped nest boxes to the sides of houses.

## What makes it special?

For a bird that lives alongside us, the House Martin is surprisingly good at keeping secrets. Given extensive research, you might think we had discovered all there is to know. However, we have failed to pinpoint exactly where they spend the winter. Millions head south in the autumn and ringed House Martins have been recorded travelling through north Africa. Yet below the Sahara, where one might expect to find British birds in their hundreds of thousands? Nothing. Well, a single recovery to be precise, in Nigeria in the 1980s. It is as if they vanish into thin air. And, strange as it may sound, that could be where they spend most of their time – high over equatorial regions, riding the thermals, sleeping on the wing at altitude and feasting on airborne insects. Advances in lightweight tracking technology may eventually enable us to follow their winter movements, but for the time being these remarkable little travellers remain among our most mysterious migrants.

## Added interest

- Cold, wet and windy weather can be extremely bad news for these aerial insectivores, especially when chicks need feeding. However, House Martins are able to sit out the hard times by dropping into a kind of torpor which reduces energy use.
- Leg warmers may have gone out of fashion in the 1980s, but never for House Martins – their legs are covered with white feathers, similar to some owls and grouse.
- It can take a pair of House Martins a fortnight to build a nest, using over 1,000 pellets of wet mud, and they readily re-use those from previous years rather than starting construction from scratch.

## Not quite a House Martin?

If your House Martin looks a little brown it could be a juvenile, but if it also has a brown chest band then it is a Sand Martin. This close relative is widespread (though less common) and a satisfying bird to identify. Up to 170,000 pairs of these migrants breed in Britain during the spring and summer. Colonies nest in holes dug into suitable sandy banks, including the sides of old gravel pits, and often hunt insects over water. Alternatively, a young Swallow could cause confusion. However, it has dark on the throat and lacks a white rump (*see Swallow*). Finally, if the sky-riding bird is dark brown above and below and has long backward-curved wings, then you are looking at a Swift (*see Swift*).

Sand Martin

# Swift

## Key identification features

- The Swift is an aerial insect-eater that has long scythe-like wings, a torpedo-shaped body, small head and forked tail.
- Apart from a pale throat, it is sooty-brown all over, though generally looks black against the backdrop of the sky.
- Groups can be seen in the summer careering low over rooftops, screaming as they go.

## Where could you find one?

These summer visitors can be seen all over Britain, though are most common in the South and East. They generally nest in colonies under the roofs of tall, old buildings in villages, towns and cities, such as churches and towers, and travel widely to feed on flying insects, often high in the sky on warm days.

## Is it easy to identify?

With pointed wings, a tapered body, forked tail and wide gape for snapping up prey, the Swift bears a strong resemblance to our other agile masters of the skies, the Swallows and martins. And yet, while it has evolved similar characteristics, it isn't even closely related – instead being placed in the same taxonomical order as Hummingbirds! The Swift is dark brown all over, while Swallows and martins have light underparts. Its wings are much longer, like backward-sweeping blades, giving it the appearance of an anchor in silhouette. The fork in the tail is not always obvious, as it can be held closed and lacks the outer feather streamers of the Swallow. In addition, the sky-riding Swift only lands in order to breed and cannot perch on rooftops or telephone wires as its tiny feet are designed simply to cling to a nest entrance.

## What makes it special?

The Swift is born to fly – and it virtually never stops. When a young bird launches itself out of the nest, it will remain airborne for an incredible three to four years before it is finally ready to build a nest and raise young. In that time it will have covered hundreds of thousands of miles, chasing the sun across Africa during the colder months before returning north over successive summers to feast on the dense 'aerial plankton' of insect life that fills our skies on long hot days. If the weather turns bad, Swifts simply fly around it. They avoid food shortages by stockpiling a ball-shaped mass of invertebrates in a throat pouch as they go, either for themselves or to feed to their young. They can also drink raindrops, mate, collect nest material and even sleep on the wing – shutting down half their brain at a time in order to keep aloft.

### How likely am I to see one?

Scan the sky. Unless you know the location of a nest site you won't see them anywhere else. They comb the airspace in a constant search for flying aphids, ants, flies, drifting spiders and moths. More than 85,000 pairs breed in the UK – arriving in May and leaving by August.

## Added interest

- Long stiff wings and tiny legs mean that a grounded bird finds it almost impossible to get airborne again. If you find one, avoid throwing it into the air, but instead, lift it aloft until it is ready to take off.
- Unusually, developing egg embryos are able to survive chilling when adults are away from the nest, while hungry young fall into a torpor to reduce their metabolism.
- Swifts may pair up young and for life, but it takes around four years before they are mature enough to breed. The oldest recorded Swift reached the age of 21.

### Not quite a Swift?

If your Swift has a noticeably forked tail and less elongated bowed-back wings, then it could be a Swallow. A decent view should reveal whether it has a light underside – which a Swift lacks. Swallows are more graceful fliers and tend to hunt flying insects lower down (*see Swallow*). If the underside is pure white and the bird has a white rump, then it is a House Martin, while the Sand Martin is light brown on the back with a band across the white chest and lacks a white rump (*see House Martin*).

House Martin

# Birds of prey

# Sparrowhawk

Female

## Key identification features

- A relatively small raptor with long, thin, yellow legs and talons which it uses to grab prey, a longish banded tail and blunt wing tips.
- The male has a blue-grey back and orangey-brown barring across the chest and underside, while the female is much larger and browner, with dark barring on the underside and a pale eyebrow.
- Both can have small white patches on the back and both have yellow eyes that gradually become more orange with age.

## Where could you find one?

This is the bird of prey most likely to visit gardens in search of food. It can be found across Britain in mainly lowland areas throughout the year and is a woodland species, building its stick nest in trees and hunting birds along the edges of woods and in clearings, along farmland hedgerows and in gardens and parks.

## Is it easy to identify?

It can be tricky to identify given that the sexes differ in size and colouring and its shape in flight alters depending on what it is doing. Nervy small birds draw attention to them by sounding the alarm with noisy calls, while the bravest may mob the passing predator in the air. When hunting, the Sparrowhawk's wings can look sharper and more falcon-like as it picks up speed, however the way it twists and turns at low level between trees and around hedges is distinctive. It has thin dark or reddish-brown barring running across its white chest and belly. Its head markings lack the dark 'moustache' of some falcons and its eyes typically have a conspicuous yellow iris. Males look grey on top, rusty orange on the face and front and have a slightly wide-eyed and baffled expression. Females are grey-brown with a light eyebrow and fiercer stare.

## What makes it special?

Sparrowhawks use the element of surprise to hunt: waiting and watching out of sight before weaving at speed along familiar flight paths and grabbing at avian prey. Arriving as if out of nowhere, these predators catch us unawares, much as they do their victims. Their appetite for small birds, including those we diligently feed in our back gardens, means they are not popular with all. Persecuted in the nineteenth century, then poisoned in the mid-twentieth century by agricultural pesticides in the food chain, our Sparrowhawks have endured tough times. Recently their population has recovered and they have spread back into former haunts. Pairs time the hatching of their eggs in late spring to coincide with an abundance of fledgling birds that make easy prey. The smaller male acts as the main provider, with the hefty nest-guarding female able to tackle larger species such as Woodpigeons and Magpies. They can certainly bring natural drama to a suburban setting - but be warned, kitchen window viewers may find some scenes upsetting…



### How likely am I to see one?

The Sparrowhawk is common and widespread, numbering over 30,000 pairs. Being a woodland species it is sparse in upland and exposed areas and the highest densities are found in central and southern Britain. They have colonised leafy towns and cities, regularly visiting gardens which provide abundant cover and prey.

## Added interest

- The size difference between males and females, (a female can weigh up to twice as much as a male) is one of the greatest of any bird species. It means that, when not cooperating to raise young, they are less likely to compete for the same kinds of prey in woodland territories.
- A hunting Sparrowhawk will not readily give up the chase and at high speed can collide with windows in pursuit of small birds.
- In falconry, the male Sparrowhawk is known as a musket, from which the firearm takes its name - the term having its origins in the Latin word 'musca' for a fly.

### Not quite a Sparrowhawk?

Our closest relative to the Sparrowhawk, the Goshawk, looks similar but is larger and much rarer. Goshawks are elusive birds of forests (particularly in Wales and southern Scotland) numbering about 400 pairs in Britain and don't hunt in gardens like Sparrowhawks.

If your Sparrowhawk has a reddish-brown back and pointed wings, which are darker at the tip, then it could be a Kestrel, a common and widespread raptor which hunts over rough grassland for small mammals (*see Kestrel*). Alternatively, if your grey-backed Sparrowhawk has a broad-chested body, a dark crown and black 'moustache' this might point to the Peregrine (*see Peregrine*), which has also taken to nesting in cities.

**Peregrine**

# Kestrel

Female

## Key identification features

■ The Kestrel is an attractive and widespread falcon, best known for its habit of hovering as it hunts.

■ The sexes differ in appearance but share a slender long-winged build, warm rusty-brown tones, black speckling and have a faint dark smudge down the front of the cheek as if their eyeliner has run in the rain.

■ Males sport an orangey back, buff underparts spotted with black, a blue-grey head and a grey tail which ends with a neat dark band. Females are more mottled brown on the back, streaked on the front and have thin bars running across the tail. Both sexes have dark wing tips visible in flight.

## Where could you find one?

Hovering over a motorway verge, hunched on a telegraph pole, hanging on the wind above a golf course, perched on a fence post… the Kestrel is not overly fussy when it comes to habitat and, as such, is one of our most widespread raptors. It will raise its young in the hollows of trees, abandoned bird nests, on cliff or building ledges and hunts in rough grassy areas large or small, rural or urban.

## Is it easy to identify?

Falcons are the fighter jets of the raptor world and can be told apart from other birds of prey by their pointed wings and lightweight build – designed for speed and agility. British falcons can be tricky to separate. The Kestrel is thinner-bodied than the Peregrine, shorter-winged than the Hobby and longer-tailed than the Merlin – which doesn't really help unless they line up in flight side-by-side. A Kestrel gives itself away by how it hunts, constantly slamming on the brakes to scan the ground below and, most characteristic of all, hovering. No other British bird hovers with the skill and stamina of the Kestrel. On average, it hovers for half a minute at a time, into a slight headwind. Kestrels also find meals by surveying promising areas from perches, and males can be identified by their toasted cinnamon backs.

## How likely am I to see one?

The Kestrel is not as abundant as it once was and has been overtaken as our most common bird of prey by the Buzzard. However, it still numbers over 45,000 pairs and is widespread year-round across a wide range of habitats - so can be encountered pretty much anywhere, at any time.

Kestrels will hunt from perches

## What makes it special?

Hovering and searching for prey

This conspicuous predator puts on a show, hovering on the breeze, holding its position, as if pinned to the sky - a hunting Kestrel is a spectacle that everyone can enjoy, even from the comfort of a passing car. It engages us with a simple question: 'Will I catch something?' Impossible to turn away as the mini-drama unfolds... While not as majestic as a Golden Eagle or as powerful as a Peregrine, a hovering Kestrel - wings beating, tail fanned, head steady - is inspiring to watch. But life at the top of the food chain is a precarious one and Kestrels have declined in number over recent decades - though the causes are far from clear. Fortunately, they can raise up to five chicks a year and will hopefully be flying high in the future.

## Added interest

- The Kestrel's excellent eyesight is sensitive to ultraviolet light, enabling it to detect the UV-reflective urine trails of voles in the grass, thus mapping its favoured prey's whereabouts more efficiently.
- The main item on the Kestrel's menu is the short-tailed vole, which it downs at the rate of at least half a dozen a day. In cities, where voles are less numerous, it turns its attention to small birds.
- Kestrels don't bother building their own nests and will instead occupy the disused nests of crows, or lay their eggs on cliff or window ledges and even in large open-fronted nest boxes.

## Not quite a Kestrel?

If the Kestrel that has swept into your garden has broad blunt-ended wings, yellow eyes, long and spindly yellow legs and reddish or brown thin barring across the chest, then you have spotted a Sparrowhawk, a specialist bird-hunter that has colonised towns and cities (see *Sparrowhawk*). Alternatively, if your Kestrel has a dark grey back and a black 'moustache', then it could be a Peregrine or a summer-visiting Hobby (see *Peregrine*). Finally, the Merlin might cause confusion - especially the female which has a brownish back. This low-level hunter, lacking in bold head markings and little bigger than a thrush, can be found in upland moorland and coastal areas, numbering around 1,100 pairs in Britain - a great sighting!

Merlin

# Peregrine

## Key identification features

■ The Peregrine is a fast-flying, bird-hunting raptor, larger than a Kestrel but smaller than a Buzzard.

■ It is a powerful broad-chested bird of prey, with wings that taper to points.

■ The back is slate-grey and the head grey-black, while the underneath is white with fine dark barring. Dark lobes below the eyes resemble a thick drooping moustache and contrast with the white cheeks and throat. The strong legs and talons are yellow and there is yellow at the base of the beak and in a thin ring around the dark eyes.

## Where could you find one?

This is a bird typically associated with wild mountain crags and sea cliffs and yet in recent years a few have taken to nesting on the ledges of high rise buildings in city centres where there is an ample supply of pigeon prey. Its strongholds are in the north and west of the country and along rocky coasts, but outside the breeding season it spreads further afield and hunts in lowland areas and over marshlands.

## Is it easy to identify?

The Peregrine, Hobby and Merlin all have grey backs and pale fronts. In a police-style line-up the Merlin would be the smallest, the Hobby next and the Peregrine the biggest of the bunch. However, out in the field it can be extremely difficult to judge size – especially as female falcons also tend to be larger than males. The Peregrine and Hobby have black hoods and moustache-like markings on white cheeks and throat. It is almost as if these flying aces are wearing those dark old-style aviator caps with flaps that come down over the ears. While the Hobby is slim and heavily streaked below and adults have a rusty-red patch under the tail, the Peregrine is heavier, paler below and has a muscle-bound chest that looks like it works out at the gym. It hunts birds by chasing them down or plunging on them from a great height.

## What makes it special?

Speed. The Peregrine is the fastest animal on the planet, reaching an estimated 200mph as it plunges onto prey. This 'stoop' involves skydiving from altitude at a flying target. It transforms in an instant, from an anchor shape soaring high in the sky to a gravity-assisted missile plummeting toward earth, accelerating as it descends. The force of impact is so great that the unsuspecting victim – anything from a thrush to a seagull – is knocked unconscious and typically caught by the Peregrine before it hits the ground. The stoop requires not only supreme aerial mastery but also sufficient strength to pull out of the dive at the right moment. Hundreds were killed during the Second World War to safeguard homing pigeons bringing messages from the battlefront and lethal effects of pesticides in the food chain sent numbers crashing. Legal protection and toxic agrochemical bans have seen this resilient species spread back from northern and western strongholds.

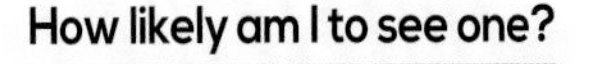

### How likely am I to see one?

There are roughly 1,500 pairs of Peregrines in Britain, mostly concentrated in the north and west and around rocky coasts. They nest at established sites and some can be viewed from manned 'watchpoints' during the spring and summer. In winter, Peregrines are more widespread and hunt over estuaries and lowland areas.

## Added interest

- Peregrines hunt birds of all sizes and over 115 avian prey species have been recorded in Britain, from tiny Blue Tits to hefty Herons. City-dwelling Peregrines even use the light of street lamps to catch birds migrating overhead at night.
- Peregrines are among the most widespread birds in the world and are found in Europe, Asia, Africa, America and Australia, with those that breed in the far north heading south for the winter.
- The Peregrine's nostrils have small bony baffles that enable it to breathe at high speeds during its near-vertical 'stoop' after prey.

## Not quite a Peregrine?

If your Peregrine looks on the slender side, with dense dark streaking on the front, long pointed wings and rufous feathering around the thighs and under the tail, it could be a Hobby. This Kestrel-sized summer migrant is an elegant and agile aerial hunter, capable of catching such aerobatic birds as Swallows and Swifts, as well as dragonflies which it eats on the wing. It is found in the southern half of Britain, hunting over lowland heaths, woodland and reedbeds and numbers about 2,800 pairs. An excellent bird to spot. Other falcons that could be confused with the Peregrine are the smaller orangey-brown Kestrel and the dynamic featherweight Merlin (*see Kestrel*).

**Hobby**

# Buzzard

## Key identification features

- The Buzzard is a large and robust bird of prey with streaky and patchily-patterned brown plumage. The tone of its colouring is highly variable: some birds are a rich dark brown while others are surprisingly pale underneath, but most are a mottled mix.

- It has a crescent brushstroke of lighter feathers across its lower chest, which may be hard to spot on creamier individuals and adults have a dark band at the trailing edge of the wings and tail. Brown patches at the front centre of the opened wings can resemble owl eyes looking down on you as it soars above.

## Where could you find one?

This is a bird of prey that lives life in the open, making it relatively easy to spot. Although most are found in the north and west, the Buzzard is widely distributed pretty much anywhere there is open ground for hunting and mature trees or cliff ledges where they can nest. They are often on the wing, circling over their territories or flying lower as they seek out prey and will also sit patiently on posts or bare branches and scan roadside verges and field edges.

## Is it easy to identify?

Birds of prey can be tricky to tell apart, being generally grey to brown and streaky underneath, with wings that are either thin and pointy or broad and raggedy. Often viewed at a distance, their subtle plumage characteristics may be hard to discern, so size, shape and behaviour become important factors in identification. On the ground the Buzzard looks stocky, muddy brown and slightly unkempt, but in the air it comes into its own, soaring like a small eagle on broad fingered wings, held slightly aloft as it catches updrafts. A dark area is noticeable just before the spread of outermost flight feathers, while its head looks small, hunched into its body and it fans its lightly barred tail wide, gliding on the thermals and filling the air with piercing cries. If you see a large, mottled-brown, wide-winged, fan-tailed raptor carving circles high in the sky then think Buzzard.

## How likely am I to see one?

Riding the thermals

Tens of thousands of pairs are scattered across Britain and they are one of our most conspicuous raptors – especially on sunny days when they ride the thermals. Countryside that combines open fields or rough ground with a scattering of woodland is likely to have Buzzards, particularly in the south west, Wales and Scotland.

## What makes it special?

Even in areas where they are relatively common, a soaring Buzzard and the sense of wildness it embodies is always a stirring sight. It may not have the majesty of an eagle, nor the awe-inspiring aura of a Peregrine, and yet the wily Buzzard is a resilient raptor that deserves admiration. Time and again its population in Britain has been knocked back. Buzzards were relentlessly persecuted by gamekeepers during the nineteenth and early twentieth centuries and driven into the remoter corners of Britain; subsequently poisoned by lethal post-war pesticides in the food chain and finally starved of prey as myxomatosis decimated the rabbit population in the 1950s. Now these protected birds of prey have been able to regain lost ground, spreading steadily from westerly strongholds into central and eastern areas. The high-flying Buzzard has become a soaraway success story.

## Added interest

- Buzzards eat small mammals such as voles and rabbits and scavenge carrion, but are not above filling up on worms and can look oddly out of place as they pace about grassy fields searching for them.
- Buzzards can also hover, briefly holding their position in the sky with heavy wingbeats.
- Females begin sitting on eggs as soon as the first is laid, which means the clutch of two to three eggs, produced over several days, hatches in sequence. In hard times the last to emerge may starve, while the larger first-born stands a better chance of surviving.

## Not quite a Buzzard?

If the large bird of prey has a fork in the tail then you've got lucky: this is a Red Kite. Once one of our rarest birds, the Red Kite has made a comeback in scattered areas through protection and reintroduction schemes and today Britain is home to more than 1,500 pairs. Long-winged and buoyant in flight, the Red Kite is noticeably reddish-brown and, viewed from below, has a tell-tale light area near the tip of the wing. Other similar birds could include the Osprey, which has long wings and is white underneath; the brown female Hen Harrier, recognisable by its white rump; the Marsh Harrier, which hunts low over reedbeds and has a pale head, and the massive Golden Eagle found in Scotland.

Red Kite

# Tawny Owl

## Key identification features

■ The Tawny Owl is our best-known owl, with its mottled brown feathering, appealing expression and classic hooting call.

■ It has a squat plump body, a big head, large black eyes and a wide, round face encircled by a hem of dark feathers and topped with high pale 'eyebrows'. Its dappled plumage is warm nut brown on the back, with some light spotting and its underparts are streaky buff.

■ In flight the wings are short, broad and rounded and it looks distinctly heavy-headed as it flaps and glides. It makes a recognisable wavering hooting sound at night.

## Where could you find one?

Widespread and fairly common, the Tawny Owl is a woodland species that can be found across Britain wherever there are a few trees, though is absent from the far north and west of Scotland, and Ireland. Pairs nest within tree hollows and other cavities and inhabit mature woods in both rural and urban areas, including churchyards, parks and leafy suburban gardens.

## Is it easy to identify?

More often heard than seen, these nocturnal hunters can be recognised by their eerie hooting, which is a familiar sound after dark in both town and country. Shakespeare gave us the memorable *tu-whit, tu-who* rendition in Love's Labour's Lost, which has stuck. However, this description actually combines two distinct elements of the owl's vocabulary: a sharp *kee-whik!* call that both sexes use to keep in touch with one another and the long quavering territorial 'song' of the male: *hooo hoo-hoo-hooooo*. If spotted at night, Tawny Owls look dark and large-headed and fly silently on rounded wings between perches. Caught in car headlights they can appear pale underneath as they cut across a road, but are nothing like as white as the ghostly barn owl.

## How likely am I to see one?

Hiding among tree foliage

This is our most common and widely distributed owl, with a population of 50,000 pairs. A nocturnal lifestyle means that individuals are hard to spot and most often encountered by chance - perhaps illuminated in car headlights. They roost in trees, tucked away within foliage close to the trunk. The owl's far-carrying hoots are most often heard in the autumn and winter as males assert their territorial rights.

## What makes it special?

There is something incredibly endearing about the Tawny Owl, from its soft rotund appearance and unhurried disposition to its open facial features and thoughtful gaze. The qualities that give it such charm are all adaptations to life as a nocturnal hunter. Soft feathers reduce noise in flight and large eyes with wide pupils enable it to see in poor light, while hidden ear openings on either side of the broad head are set at differing heights to help pinpoint sound direction. It relies on acute hearing, more than eyesight, to catch prey, listening out from perches before swooping down silently on small mammals, roosting birds or even frogs. Although the Tawny Owl's night vision is impressive, aided by an abundance of light-sensitive receptors in the retina, it cannot see in pitch blackness. An excellent spatial memory of the layout of its lifelong woodland territory helps it get about safety in the dark.

## Added interest

- In wet weather, when the rustlings of small rodent prey are obscured by the noise of wind and rain, Tawny Owls will pace about on the ground eating earthworms. They have even been known to catch fish in shallow water.
- Fluffy young owlets clamber out of the nest before they can fly, but remain on neighbouring branches where they are fed by their parents.
- Tawny Owls normally eat prey whole and regurgitate indigestible fur, feathers and bones as pellets, which can be examined to shed light on their diets.

## Not quite a Tawny Owl?

If your owl has large tufts above the ears and orange eyes, then it is a Long-eared Owl. Well spotted! Several thousand pairs of this secretive, slender and streaky nocturnal species breed in scattered areas of Britain, often in conifer plantations. Hunting low over rough ground in daylight with yellow staring eyes and long black-tipped wings that are pale underneath? Then you are looking at a Short-eared Owl (*see Barn Owl page*). Finally, you might come across a small, plump brown owl, little bigger than a thrush, hunting by day from a perch out in the open: the Little Owl. This lowland farmland species, with frowning pale 'eyebrows' and a distinctive head-bobbing habit has an overall population of more than 5,500 pairs.

Long-eared Owl

# Barn Owl

## Key identification features

- The Barn Owl is a distinctive bird, its back the colour of golden syrup, mottled with grey and freckled with dark spots, while the underparts and underwings are pure white.
- It has black eyes set within a heart-shaped face, fairly long round-ended wings and can be seen hunting by day. Snowy white and floaty in flight, it makes a screeching sound, nothing like the hooting we normally associate with owls.

## Where could you find one

Thinly spread across most of Britain, the Barn Owl can be found year-round in open lowland areas which offer sheltered places to roost and rough grassland containing plentiful small mammal prey. They nest in tree hollows, barns and derelict buildings and hunting grounds include meadows, field edges, road verges and grassy stretches beside hedgerows, railways, rivers and marshes.

## Is it easy to identify?

While the nocturnal hooting Tawny Owl is typically heard but not seen, it is the opposite with the less common and more rural Barn Owl. This dedicated rodent-catcher frequently hunts at dawn and dusk in full view, especially when it has young to feed or during the winter. They can be spotted perched on fence posts or flying low over fields, methodically covering ground with buoyant flight. Lightweight and broad-winged they stall and hover, dark eyes fixed on the grass beneath, before plunging down with long legs extended to grab prey in their sharp talons. Their underparts and underwings are remarkably white and, in good light, the speckled golden back is visible. Caught in car headlights at night their ghostly apparition is much paler than that of our other owls, which have darker backs and streaking on the front.

## What makes it special?

In a recent vote to choose Britain's national bird the Barn Owl came second after the Robin. For many people, it is their favourite species. Large front-facing eyes and wide facial features that mirror our own may partly explain its appeal, but it also has an ethereal quality that sets it apart. In the past, this creature of the night was, perhaps understandably, the source of much superstition. While the owl retains its sense of mystery, magic and the macabre, it has also acquired new meaning as a symbol of a healthy countryside. Barn Owl numbers plummeted by more than 70 per cent during the twentieth century and its plight helped inspire campaigns to provide space for nature in intensively managed landscapes, including the provision of nest boxes. This much-loved champion of the conservation cause has rewarded efforts by gradually increasing in number – sightings of these special birds come with an added feel-good factor.



### How likely am I to see one?

There are more than 4,000 pairs breeding in Britain. During the summer, when the nights are short and there are hungry chicks to feed, Barn Owls can be seen hunting during the day – in particular at dawn and dusk. In the colder months numbers are boosted by winter visitors from the Continent.

## Added interest

- Barn Owls can catch prey without seeing it. The shape of the face funnels sound towards highly sensitive ears positioned asymmetrically at the outer edge – the slight height difference between left and right ear openings enabling it to work out the direction of noises more precisely.
- A comb-like fringe at the front of the wing feathers and a soft trailing edge allows a hunting Barn Owl to fly silently and hear small mammals more clearly, while avoiding alerting prey to its presence.
- A nesting female lays an egg every other day, but begins incubating from the outset, so the five or so chicks hatch in sequence – with the youngest and smallest most likely to starve if food is short.

### Not quite a Barn Owl?

The species most likely to be confused with the Barn Owl is the Short-eared Owl, which can also be seen hunting over open country in daylight. In flight, the Short-eared Owl's tapering brown body and flat face look a bit like a horizontal ice-cream cone, while its slender dark-tipped wings are light underneath and marked with a black crescent at the bend. The yellow eyes are surrounded by a smudge of mascara black and stand out even at a distance. Scarce in the summer, with around 1,500 pairs breeding mainly on uplands and northern moorland, it is more widespread during the winter when an influx of Continental Short-eared Owls boosts numbers.

Short-eared Owl

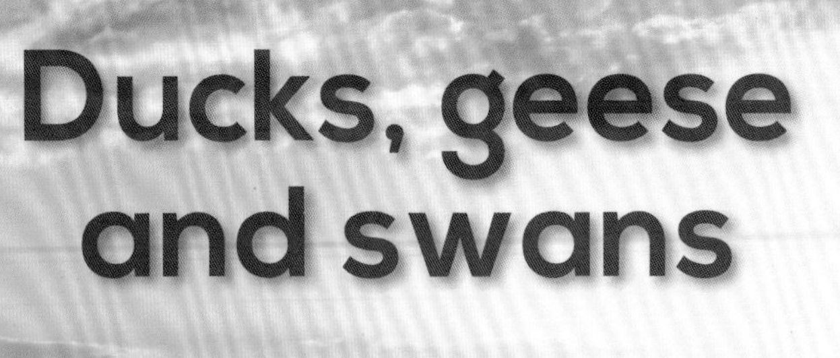
Ducks, geese
and swans

# Mute Swan

## Key identification features

■ The Mute Swan is a familiar, graceful bird of lakes and rivers, with white plumage and an orangey-red beak which has a black base and knob on top – a bump that is larger in the bigger males.

■ Young birds are grey-brown, while the adults are pure white, though the head and neck feathers, submerged when feeding on aquatic weed, may be stained rusty-gold by tannins and iron in the water.

■ In flight, Mute Swans hold their necks out straight and the beating wings make a rhythmic throbbing sound. Pairs defending a nest raise fluffed-up wings over their backs and paddle forward with double-footed surges.

## Where could you find one?

Widespread in lowland areas year-round, the Mute Swan can be found across most of Britain – in particular central and southern areas – on lakes, well-vegetated reservoirs, canals, wetlands and slow-flowing rivers, as well as around estuaries and sheltered coasts. As one of the world's heaviest flying birds they need space for a running take-off and they land on water, as a rule, to avoid injury - though they may wander from lakes and rivers to graze in neighbouring fields.

## Is it easy to identify?

Massive, white and shaped like a swan – easy. Except that, just to confuse matters, two other swans can be seen in Britain: the Whooper Swan and the Bewick's Swan. These winter visitors from Iceland and Siberia have yellow and black beaks, rather than orangey-red and black. The Whooper Swan is much the same size as a Mute Swan, with a wedge-shaped head and bill and a generous amount of yellow on the beak extends well forward in a pointed patch. The Bewick's Swan, on the other hand, is smaller, with a modest blunt-ended patch of yellow on the beak. Both of these wild swans are more vocal than the Mute Swan (which, despite its name, does make various grunting and hissing sounds). They also tend to hold their necks up straight when swimming, rather than adopting the elegant 'S'-shaped curve of the Mute Swan.

## How likely am I to see one?

More common in winter

The Mute Swan is our most common swan and resident year-round. Roughly 6,000 pairs breed in Britain, though in colder months this number reaches tens of thousands when migrants from northern Europe join winter gatherings. Gradually replacing their feathers in the late summer and autumn they temporarily become flightless for several weeks.

## What makes it special?

Majestic, dignified and imposing. While excavated remains show they have been present in Britain for thousands of years, the origins of our present day population are complicated by a history of transportation, trade and domestication. That we can enjoy seeing Mute Swans in Britain at all is largely down to the fact that they were considered the ultimate high-status table bird at medieval feasts and strictly protected by nobility from being hunted to extinction. Rights of ownership, bestowed by royalty, were signified with individual bill markings (of which 900 were recognised during the reign of Elizabeth I) and wings were clipped or pinioned to prevent valuable flocks from roaming. Unmarked birds on open water remain the property of the crown, though the Queen only exercises her claim on the Thames. Protected from harm by new laws, including a ban on the use of toxic lead fishing weights, our magnificent Mute Swans no longer live in bondage to the aristocracy and are free to spread their wings, rulers over their own watery domains.

## Added interest

- Mute Swans pair for life, which is typically about ten years, though will find a new mate if their partner dies.
- The warning that an angry swan can break your arm with a single wingbeat is exaggerated, however these powerful birds are certainly capable of causing injury to the young or frail.
- Rights of ownership are still exercised on the Thames by the Crown and two historic corporations in the ceremonial practice of 'swan-upping', where birds are rounded up and entitlement to the year's cygnets signified using leg rings, rather than traditional nicks carved into the bill.

## Not quite a Mute Swan?

If your swan has a yellow beak rather than the typical orangey-red then it is one of the wild winter swans that visit wetlands, flooded fields and estuaries every year, escaping freezing conditions further north (*see Is it easy to identify*?). Around 15,000 Whooper Swans, with their doorstop-shaped beaks, come from Iceland and flocks are mainly seen in Scotland, northern England and East Anglia. The smaller Bewick's Swan, which has more black on the bill, breeds in Siberia and roughly 7,000 winter in scattered areas, mainly in England.

Whooper Swan

# Canada Goose

## Key identification features

- The Canada Goose is a large and handsome brown goose with a black head and neck and a distinctive white patch that runs from the side of the face under the chin.
- The neck is long and slim and the heavy body is pale on the chest and darker brown on the back, finely barred with grey, while the plumage above and beneath the base of the dark tail is white. They make a honking call, typically in flight.

## Where could you find one?

From park lakes, reservoirs and flooded gravel pits to canals, wetlands and estuaries, the common Canada Goose can be found over much of lowland Britain, particularly central and southern England. They nest near water, forage in parks, grassy fields and on arable land, and typically roost on or beside open water.

## Is it easy to identify?

The Canada Goose stands out from the crowd, not only on account of its size among other geese and ducks, but also its neat head markings. The black of the bill, head and neck, neatly ruled off from the rest of the body, looks as if the bird has been dipping deep into a pool of ink to feed, while the broad 'chinstrap' of white against the dark face is obvious even at a distance. This familiar bird of park lakes can become tame enough to take food from the hand and is a common sight grazing on mown areas. Although its robust grey-brown waddling body is very goose-like, the contrasting neck and head can look slender and elegant when outstretched. And when it comes to identification, it helps that both sexes look the same and that goslings rapidly grow up to resemble a duller version of their parents.

## What makes it special?

As the name suggests, the Canada Goose is a species of the North American continent and not a native British bird, though it has been here for so long that it is now a well-established resident. First introduced to the waterfowl collection of King Charles II at St James Park, London, in 1665, it became a popular addition to lakes at stately homes, admired for its size, confidence and striking black-and-white markings. It was a couple of hundred years before these ornamental geese began to spread their wings and breed in the wild, successfully conquering much of lowland Britain. While counterparts in Canada and the United States are migratory, ours have lost the wanderlust and largely stay put – especially where the living is easy. It is an undeniably handsome bird with bags of character and in urban areas, where it can be particularly tame, it provides an opportunity for young and old to enjoy close contact with nature.

### How likely am I to see one?

In the 1950s there were fewer than 4,000 living in Britain, while today around 62,000 pairs breed and the winter population approaches 200,000 individuals. It can be abundant at town and city lakes where it has learnt to take advantage of food hand-outs and the relative warmth and safety of urban living compared with rural areas.

## Added interest

- While Canada Geese in Britain were originally introduced, migrating wild birds from North America do occasionally take a wrong turn and wind up on our northerly and westerly shores in winter.
- Canada Geese are territorial when breeding, though will nest in colonies and gregarious goslings from neighbouring pairs may join together in crèches guarded over by adult birds.
- When flying long distances, flocks adopt a 'V' formation to improve efficiency, sharing uplift and reducing wind resistance for all but the front position, which is rotated between birds.

### Not quite a Canada Goose?

The side-swipe of white on the black head is the signature mark of the Canada Goose, however there are a couple of other geese which could cause confusion. The Barnacle Goose is smaller, greyer, with the black on head and neck extending down onto the chest and has white over the front of the face. The Brent Goose is small, sooty-grey and has an entirely black face, along with the neck and chest. The only white near the head is a thin brush-stroke across the upper neck of adults. In flight they look very dark, with white at the rear end. Tens of thousands of both Barnacle and Brent Geese are present in Britain during the colder months.

Barnacle Goose

# Greylag Goose

## Key identification features

- The Greylag Goose is a hefty greyish-brown bird, with a heavy body, thick neck and a robust wedge-shaped beak. Its dark back is barred with the thin lines of pale feather fringes, the underside is pale and the base of the tail is white, both above and beneath.
- It waddles on stocky pinkish legs, has furrowed lines in the plumage on the sides of the neck and the thickset bill is orange.
- In flight, it is noticeably light grey across the front of the wing and on the rump and it makes the typical honking sounds of a farmyard goose.

## Where could you find one?

Lakes with plenty of bankside vegetation and nearby fields in which to graze provide the ideal breeding habitat for this widespread resident species – especially if the open water has islands where they can nest on the ground in safety, out of reach of predators such as foxes. Greylag geese can be found on reservoirs, flooded gravel pits, park lakes, lochs, marshes and estuaries and, in the winter, flocks forage on pasture and arable land.

## Is it easy to identify?

This gregarious, vocal bird can sound a bit like an old-fashioned car horn and flocks in flight may form lines or 'V' formations – always an uplifting spectacle. From late autumn to spring, Britain is visited by similar-looking grey-brown wild geese that breed in the Arctic and spend the colder months here; telling the species apart can be tricky given that their body plumage is virtually identical (*see Not quite a Greylag Goose?*). The Greylag is the beefiest of the bunch and its orange doorstop beak is a key distinguishing feature. This sturdy triangular bill is visible even in flight, when the pale rump, silvery-grey front to the upper wing and two-tone contrasts in the underwing help distinguish it from other lookalike geese.

## How likely am I to see one?

Birds graze in fields

Widespread and common year-round, breeding Greylag Geese number over 45,000 pairs and are most easily spotted on freshwater lakes in lowland areas. The population is boosted between autumn and spring by Icelandic flocks, totalling over 85,000 birds, which come here to avoid freezing winter conditions further north.

## What makes it special?

The Greylag, our only native breeding goose, is a bird with a split personality: on the one hand a bold and brash resident of urban lakes and open water and on the other, a wary wanderer of the north. While closely associated with humankind for millennia, the species still endures in remote areas and wild flocks make for an inspiring sight. Once widespread, Greylag Geese were hard hit by habitat loss and hunting, becoming extinct as breeding birds across England in the 1800s, with remaining populations restricted to north and west Scotland. In the twentieth century, birds were reared and reintroduced to Wales, southern Scotland and England (using eggs taken from Scottish strongholds) and have since spread widely. Today our Greylags also include a mixture of migrating birds from Iceland that spend the winter months in the north of Britain. At home in settings ranging from park lake to windswept Arctic tundra, the Greylag Goose is a tough and adaptable species.



## Added interest

- Greylag Geese occasionally mate with other species, such as Canada and Barnacle Geese, creating hybrids with a mixture of plumages. There are even records of half-Swan, half-Greylag offspring, nicknamed 'swoose'.
- Male Greylag Geese take their nest-defending duties seriously and celebrate repelling any enemy by returning triumphantly to their mate and showing off with a noisy display.
- The Greylag Goose was domesticated more than 3,000 years ago and revered by ancient cultures which associated it with gods of the sun and fertility.

## Not quite a Greylag Goose?

In the autumn, other species of grey-brown goose with similar body plumage arrive for the winter, roosting at traditional scattered wetland sites and foraging in fields. Fortunately the names of these long-distance travellers help with identification. The White-fronted Goose, from Greenland and Siberia, has a white front to the face, orange legs and adults have ragged black bars on the belly. Over 15,000 visit south and east England, the Severn estuary and parts of Scotland and Ireland. The Pink-footed Goose has pink legs and feet, as well as a chocolate brown head and a short bill with a pinkish band across the middle. Over 300,000 come here from Arctic breeding grounds, concentrating in East Anglia, the north-west and Scotland.

White-fronted Goose

# Shelduck

## Key identification features

- The Shelduck is a conspicuous coastal bird, roughly halfway between a duck and a goose in size.
- Both sexes are boldly marked, with a greenish-black head, red beak, pink feet, black shoulders and a thick chestnut belt that runs around the lower breast and back of the mostly white body. The male has a red knob where the top of the bill meets the forehead.
- In flight, the outer feathers of the Shelduck's white wings are black and a dark stripe is visible running down the belly.

## Where could you find one

Shallow tidal mudflats and estuaries, rich in invertebrate life, provide the perfect year-round habitat for Shelducks, which glean small aquatic snails, worms and crustaceans from the ooze. They breed along sheltered coasts, nesting in concealed sites such as abandoned rabbit burrows, bushes or barns. Once the eggs have hatched, both parents lead the ducklings overland to a territorial stretch of water.

## Is it easy to identify?

A Shelduck always looks like a Shelduck whether winter or summer, male or female – which makes year-round identification reassuringly straightforward. Even from a distance these large and neatly-patterned birds stand out, their white plumage catching the eye and contrasting with clean blocks of colour: glossy dark green on the head, inky black on the shoulders and wing tips and ginger brown in a broad band around the middle like a life-ring. In the air, Shelducks look quite black and white; in poor light, when colours are dark and indistinct, the basic alternating plumage segments of head, chest, breast band and flanks are unique (the term 'sheld' means pied or variegated). They can often be seen sifting the surface of wet mud for food and may upend in shallow water to reach molluscs and crustaceans.

## How likely am I to see one?

During the spring and summer breeding season, some 15,000 pairs nest around our coast (favouring sheltered tidal stretches) and are fairly easy to spot. In the winter, migrants from mainland Europe swell our resident population to tens of thousands of birds and large feeding flocks can be seen on muddy estuaries.

## What makes it special?

The Shelduck is one of the most common species of wild duck breeding in Britain. It has a complicated family life, which involves seemingly dedicated parents deserting their ducklings in order to moult. Shelducks replace their feathers in late summer and, in typical wildfowl fashion, they shed all the flight feathers at once, so are rendered flightless for four weeks while the new set grows. Because this is such a vulnerable time, adults gather in flocks at safe coastal locations. While some head for traditional sites in Britain to moult, the majority join vast congregations on the tidal flats of the shallow Wadden Sea off north-west Germany. That means abandoning their young, which join together in groups guarded by an adult or two until they are old enough to fend for themselves. It is a bit like the parents dumping their offspring in summer camp before flying off on a European break.

## Added interest

- In general, female ducks are the dowdier of the sexes, with plainer plumage that enables them to blend in when sitting on an exposed nest. Not so female Shelducks, which have to nest in holes and thick vegetation in order to keep out of sight.
- One of the Shelduck's main foods is a tiny mud snail which it hoovers up in vast quantities, eating hundreds and even thousands at a time.
- While Shelducks typically lay eight to ten eggs, some may end up with much larger clutches as females without nests occasionally lay in a neighbour's when it is left unattended.

## Not quite a Shelduck?

If your Shelduck has the standard greenish-black head and white chest, but a broad, dark beak and rufous-brown flanks, instead of a band around the lower breast, then you are looking at a male Shoveler. Generally a species of shallow freshwater lakes and marshes, the Shoveler is a good find in the summer, as only about 1,000 pairs breed in Britain, particularly in southern and eastern areas – though there are many more in the colder months as migrants fly here for the winter. The female is mottled brown and looks much like a female Mallard (*see Mallard*). Both Shoveler sexes have a distinctive large, wide bill that they use to filter invertebrate food from the water.

**Shoveler**

# Eider

Female

## Key identification features

■ The Eider is a handsome and hefty sea duck with a distinctive wedge-shaped beak.

■ Males are white above, black below and have a black cap. In flight, generally strung out in groups low to the sea, they look pied and heavy bodied with a black belly. Close views of drakes reveal a pink flush on the breast and green around the back of the head.

■ The female is mottled brown, much like a female mallard, but has the Eider's characteristic long triangular bill.

## Where could you find one?

This is a marine duck of coastal waters, at home riding the waves along rocky shores or diving in tidal bays after shellfish and crustaceans – only very rarely seen on freshwater bodies inland. Eiders breed in Scotland, northern England and Northern Ireland, nesting close to the sea among rocks, vegetation and ruined buildings in sheltered spots or on islands. In winter they are joined by visiting migrants from northern and eastern Europe and can be seen all around our coast.

## Is it easy to identify?

One thing stands out about the Eider: its beak. A sturdy wedge, it gives the whole head a geometrical quality. The face tapers into it, with cheek feathers extending over the bill base and the dark crown of the male runs along the top like the brim of a baseball cap. Even at a distance or in poor light, the Eider's distinguished doorstop profile sets it apart. It is also a large and powerful species, thick-necked and broad-bodied.

Seen well, the eccentrically-marked drake is unmistakable, sporting pure white upperparts, black sides, cap, belly and tail, a rosy blush across the chest and a smudge of celery-green over the back of the head. When courting during the winter, they throw their heads back and make an endearing cooing sound. The female is plain and well-camouflaged and would be hard to identify were it not for the long-faced appearance.

## What makes it special?

Being diving ducks of cold seas, they need a thick insulating layer of body feathers to keep warm. This soft down has long been prized for its heat-retaining properties and has been used for centuries to stuff quilts, duvets and pillows. You don't need to hunt Eiders to obtain their feathers – they do all the selecting, plucking and stockpiling themselves. Breeding females keep their clutch of eggs snug by lining the ground nest with a deep layer of their own greyish down feathers. This lightweight down has been traditionally harvested, though while protected colonies still support a small sustainable industry in countries such as Iceland, the valuable bedding material has been superseded by cheaper alternatives. Eiders are abundant across the Arctic, diving among icebergs or toughing out storms along rocky coasts. Once a barrel-chested, hard-flapping Eider gets airborne it is one of the fastest birds in the world in sustained level flight, having been recorded at just over 47mph.

### How likely am I to see one?

The Eider is one of our most common breeding ducks, with roughly 30,000 pairs nesting along coastlines in the northern half of Britain. During the colder months the population, swelled by over-wintering visitors, is spread more widely around our shores, extending as far as southern England.

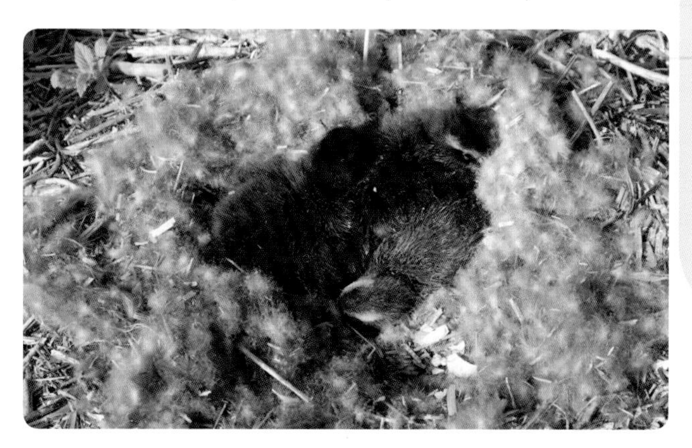

## Added interest

- In Northumberland, the Eider is known as the 'Cuddy duck' after Saint Cuthbert, who lived among them as a religious hermit on the Farne Islands in the seventh century and forbade anyone from harming the colony – which still exists today.
- Young males take until their fourth year to fully acquire smart adult plumage and start breeding, though may live to a good age – the oldest recorded individual died aged 35.
- Eider ducks sometimes nest within noisy colonies of Arctic Terns that aggressively defend their breeding grounds, offering added protection against predators such as foxes and gulls.

### Not quite an Eider?

The male Eider is one of a kind, with his inky belly, white upperparts and black cap that resembles a beret pulled a little too far down over the eyes. Shelduck may cause confusion as they also have a mixed plumage of dark and white segments, though these similarly coastal birds have a chestnut breast band and red beak (*see Shelduck*). During their summer moult, male Eider are much darker and messier looking. If your mottled-brown female Eider looks on the small side, lacks a telltale wedge-shaped beak and has a flash of purple-blue on the inner wing then it could be a Mallard (*see Mallard*). And if it has a broad, flattened bill then it could be a female Shoveler (*see Shelduck*).

**Shelduck**

# Mallard

##  Key identification features

- The Mallard is a handsome and familiar duck of lakes, rivers and ponds.
- The male's iridescent dark green head and neck has a lustrous quality, as if dipped in metallic paint, and is divided with jigsaw-set simplicity from its chestnut breast by a thin white collar. It has a yellow bill, orange feet, a grey and brown body and a black and white rear end which sports up-curled black central tail feathers.
- The female is mottled brown and has an orange-brown bill. Both sexes have a shiny purple-blue patch, fringed with white, on the inner wing, which is conspicuous in flight.

##  Where could you find one?

For Mallards, just add water. They can be found year-round pretty much anywhere wet, ranging from park lakes, canals and village ponds to reservoirs, slow-flowing rivers, marshes and estuaries across the British Isles.

##  Is it easy to identify?

Mallards are the standard duck with which most people are well acquainted. Smart-looking males, with their polished green heads, stand out in the crowd. The streaky brown females are also recognisable, especially as they are the ones that deliver the loud quacking sound we associate with ducks – a raucous call which can be repeated in rapid succession like a grating laugh. However, they may also be confused with other female ducks and the white-edged blue patch on the inner wing is a key distinguishing feature. In late summer after the breeding season, male Mallards lose their bright colours as they moult their plumage. The replacement of wing feathers renders them temporarily flightless and they adopt a low-key camouflaged appearance similar to females, which makes them trickier to identify. Despite this, they can still be told apart from females during this so-called 'eclipse' plumage phase by their yellow bill colour.

## How likely am I to see one?

Upending to find food

The Mallard is our most common duck, with up to 150,000 pairs breeding across Britain. Numbers increase several-fold in the winter as birds from Iceland and northern Europe retreat from freezing conditions and come here to paddle in our relatively frost-free lakes and estuaries.

## What makes it special?

Few British birds enable people of all ages to enjoy such close contact with nature in built-up areas as the Mallard. The species' remarkable tolerance of humans has enabled this attractive and adaptable duck to live alongside us, giving pleasure to countless park-goers – in particular, the delightful sight of a female with her brood of fluffy ducklings in tow. Yet while the Mallard has become unusually tame on urban lakes and canals, those living in the wider countryside can be surprisingly wary, which is a wise policy as they are widely hunted. A truly wild Mallard chanced across on an isolated loch or dabbling at the edge of a secluded mire remains a special sight.

## Added interest

- The female Mallard lays an impressive clutch of around a dozen eggs – the equivalent of half her body weight – and relies on a concealed nest site and camouflage plumage to avoid detection as she incubates them for nearly a month.
- Males play no part in rearing young and, for a period during the summer, unattached gangs relentlessly harass and forcibly mate with lone females, which can result in drownings.
- Mallards have taste buds positioned on the inside of their beaks, allowing them to assess the palatability of foods simply by picking them up.

## Not quite a Mallard?

Mallards are the ancestors of an assortment of domestic duck breeds, ranging from the pure white Aylesbury to the upright skittle-shaped Indian Runner Duck. There are a few wild duck species with which Mallards can be confused. The dark bottle-green head of males is shared by the Shelduck and male Shoveler (*see Shelduck*) as well as the slender Goosander, which has a thin red bill. Female Mallards look very similar to a number of brown female ducks, including the Gadwall, which has a white inner-wing patch; the broad-billed Shoveler and the Eider with its wedge-shaped beak (*see Eider*), as well as the female Pintail and the small Teal (*see Wigeon*).

Male Goosander

# Wigeon

Female

##  Key identification features

■ The Wigeon is a delightful medium-sized duck with a small beak and rather gentle appearance.

■ The male is particularly handsome, having a rich chestnut head with a creamy-yellow smudge running down the forehead and a neat grey body that is pinkish across the breast. The belly is white, the tail black and pointed and a white wing patch is visible in flight and across the side of the folded wings.

■ The mottled reddish-brown female lacks the showy plumage of the male, but shares the round-headed shape, black-tipped grey bill and white belly. Males have a distinctive whistling call.

##  Where could you find one?

Only a few hundred pairs breed in Britain, beside lakes in Scotland and northern England. However, in the autumn vast numbers arrive from Iceland, Scandinavia and Russia to spend the winter months here. Their diet includes grass, green seaweed and eelgrass: flocks feed on estuaries, saltmarshes and coastal mudflats, as well as inland wetlands, reservoirs and flooded gravel pits. They can be seen in suitable habitat across most of the country in winter, dabbling at the water's edge or grazing in fields.

##  Is it easy to identify?

The grey-suited, pink-waistcoated male looks every bit the dapper gentleman, with a distinctive blond streak running from the top of the beak to the crown, as if his forehead has caught the sunlight. In flight, a prominent splash of white across the inner wing is noticeable and on the ground this shows as a visible stripe along the side. Males also draw attention to themselves with whistling *whee-ooo, whee-ooo* calls. The rusty-brown-and-grey females have rather plain mottled plumage which enables them to keep a low profile when sitting on eggs. However, they share the same shape and little grass-nipping bill as the males, as well as dark legs and a white belly. During the annual moult after breeding, male Wigeons lose their bright colours and become much more dowdy and inconspicuous – a sensible precaution as they are rendered temporarily flightless in the late summer while growing replacement wing feathers.

## How likely am I to see one?

A handsome male

Mid-winter is the best time to see Wigeon, when large numbers can be found across much of Britain on wetlands and mudflats. Some 450,000 birds spend October to March here and many estuaries hold decent numbers. During the spring and summer our population dwindles to 300-400 resident breeding pairs, found at freshwater sites in Scotland and northern England.

## What makes it special?

The male Wigeon, sporting finely-stitched grey feathering and a flush of pink across the breast, is arguably one of our most attractive species of wildfowl. In the autumn, when these impeccably-dressed drakes arrive with females in their hundreds of thousands from Arctic breeding grounds, swelling our own small resident population, they come with a characterful whistle. This endearing and un-duck-like note, that rises and falls like a short squeeze on a plastic bath toy, is among the most evocative sounds of wetlands in winter. It penetrates chilly mists and carries far over open water, expanses of tidal mud and damp pasture, revealing the whereabouts of chatty and gregarious flocks feeding on estuaries, marshes and fields. Wigeon are a heart-warming sight (and sound) on a cold day.

## Added interest

- The name Wigeon can be traced back to the early-1500s and is thought to derive from the bird's whistling call, which also generated alternative terms such as 'whewer' and 'whistler'.
- The Wigeon's low calorie vegetarian diet means it must spend much of the day feeding to meet its energy requirements - even if that means snacking after dark.
- Winter concentrations of Wigeon in Britain are of international importance and some wetlands, such as the Ribble Estuary in Lancashire and Ouse Washes in Cambridgeshire, can have tens of thousands present at a time.

## Not quite a Wigeon?

If your Wigeon is lacking the light-yellow forehead and has a black breast, not pinkish-grey, then it is a Pochard. This diving duck of freshwater lakes and reservoirs, nests in scattered areas. Flocks may be made up of mainly males and they can be spotted on park lakes. Alternatively, the Teal could cause confusion. This nervy, fast-flying little duck has a grey body with a thin white stripe along the side and chestnut head. However, it also has a bottle-green patch on the face running back from around each eye, pale yellow feathering under the tail and a white-bordered bright green patch visible at the rear of the open wing. Teal are most abundant in the winter, totalling more than 200,000 birds.

Pochard

# Tufted Duck

Female

## Key identification features

- The Tufted Duck is a charming medium-sized duck, named for the wisp of feathers that droop down over the back of the head.
- The male is glossy black and has a broad rectangle of white on the sides and across the belly, a black-tipped blue-grey beak and a long tuft.
- The female is chocolate brown, darker on the upperside, lighter beneath and has a short tuft. Both have bright yellow eyes and, in flight, a white stripe runs along the back half of the wings.

## Where could you find one?

Resident year-round, the Tufted Duck breeds next to open water on well-vegetated banks and islands. It nests, often in loose colonies, beside lakes, reservoirs, flooded gravel pits and wide sluggish rivers: it can be found in lowland areas across England and much of Scotland, though is scarce in Wales. In the winter this gregarious bird gathers on similar inland stretches of water in much the same areas. Tufted Ducks are also familiar inhabitants of urban park lakes, where they can become quite tame.

## Is it easy to identify?

The Tufted Duck is one of a group of species that dive to find food. With a forward leap they plunge down several metres on forays lasting a quarter of a minute or so, snatching freshwater mussels, crustaceans and invertebrates from the lake bed. When they pop to the surface, with beads of water running off their well-oiled plumage, they look so buoyant it seems a miracle they can get down beneath the surface at all. Their black-and-white appearance, short-necked profile and constant diving set them apart on freshwater lakes from other species, along with the slicked-back tuft of head feathers – which is shorter, but still visible, on the dowdier females. Golden-yellow eyes, like sequins pinned onto the soft glossy head, are also distinctive. In the summer, males lose their crisp contrasting plumage for a few weeks while they are moulting and growing new feathers.

## How likely am I to see one?

Visible white wing stripes

More than 16,000 pairs nest across Britain in the breeding season, with the greatest concentrations in central and eastern areas. Over the colder months, visitors from Iceland and northern Europe push up numbers to 120,000 individuals. The largest over-wintering gatherings can be found in Northern Ireland and groups on open water may mix with other species.

## What makes it special?

In the early 1800s, this characterful diving duck was one of our rarest breeding birds. The small numbers that regularly visited Britain in the winter from Iceland, Scandinavia and northern Europe, only began to settle here from the middle of the nineteenth century. Since then it has spread far and wide in suitable watery habitats. It has also been helped by an invader from the Caspian Sea: the zebra mussel, which was accidentally introduced to Britain around the same time as Tufted Ducks started to colonise. These stripy little freshwater bivalves can be a nuisance, damaging ecosystems and clogging up pipes, but they are much loved by the ducks, which wolf them down in vast quantities. Tufted Ducks are a perky-looking species and provide added entertainment when they dive close by, enabling one to follow their dark shapes in the murky water, webbed feet pumping behind as they scour the muddy bottom for food.

## Added interest

- In the winter, Tufted Ducks can eat over 1.5kg of small freshwater mussels every day, hoovering them up with their broad beaks and swallowing them whole, before they are crushed in the gizzard.
- To avoid become waterlogged, they waterproof their plumage with oil from a special preen gland, situated at the base of the tail, using their head and bill to rub it over their feathers.
- The distinguished back-combed tuft on the heads of males is longest in the winter and spring breeding season, but much shorter during the summer period while they are moulting.

## Not quite a Tufted Duck?

If your Tufted Duck is lacking a crest and its back is grey, instead of black, then it is a Scaup. This very similar-looking diving duck is a winter visitor and flocks are mainly seen in sheltered coastal areas. It derives its unusual name from the Scottish term for the mussel beds on which it feeds and is twenty times scarcer than the Tufted Duck. The brown female Scaup is much like a female Tufted Duck, however while female Tufted Ducks may occasionally have a little white above the base of the bill, the female Scaup has a noticeably large patch in front of the eyes and no hint of a crest.

Scaup

# Other lake and river species

# Kingfisher

Female

## Key identification features

- The Kingfisher is a stunning-looking bird with orange underparts, greeny-blue wings and head, cheek patches of orange and white, and an iridescent vivid light blue streak running down the back and rump.
- It is smaller than many expect (roughly Sparrow-sized) and has a dumpy little body, short tail and rounded wings, a proportionately large head and long dagger-like bill.
- Females have a lipstick-smudge of orange-red at the base of the black bill.

## Where could you find one?

As the name suggests, the Kingfisher is a supreme fish-catcher, but its small size means it can only tackle little fish (such as minnows and sticklebacks) relatively near the surface, relying on excellent vision. It requires sheltered, clear and shallow water and is found beside slow-moving rivers, tree-fringed tributaries, canals and still water. In the winter it may also turn up at estuaries and in coastal areas.

## Is it easy to identify?

Despite startling plumage of breath-taking intensity, the Kingfisher is surprisingly easy to miss. They typically hunt from a waterside perch where dappled light beneath riverside foliage dulls iridescent colours and their still, small form patiently surveying the water beneath may be overlooked. If you are lucky enough to see a Kingfisher at rest then it is pretty unmistakable, with vivid blue and orange plumage and a distinctive pointed shape – all beak and head on a stubby body. They are far more likely to be spotted in flight, following a watercourse on short whirring wings, revealing their astonishing azure backs as they whizz by at speed and frequently drawing attention to themselves with short, high-pitched calls. However, sightings can be tantalisingly brief, leaving just an impression of electric brilliance etched in the mind.

## How likely am I to see one?

A patient waterside hunter

There are several thousand pairs of Kingfishers in Britain. They aggressively defend territorial stretches of water at least a kilometre in length and a walk beside calm, slow-moving water, a lake or flooded gravel pit offers the chance of an encounter. However, they are not easy to get close to and more likely to be observed perched at a distance.

## What makes it special?

The Kingfisher is an iconic bird, the darling of calendar picture editors and garden ornament designers. However, much like the Cuckoo, it is an elusive ornithological celebrity, well known but seldom spotted and always a privilege to glimpse – even if fleetingly. Tropical radiance, endearing size, a photogenic perching pose and plunge-diving abilities all add to the appeal of this one of a kind British species. More than simply attractive, it is a skilful master of the art of fishing which rightly earns our admiration. To raise a brood of, on average, six young, a pair must catch more than 100 fish a day. The presence of Kingfishers on a river indicates it is in good health, with populations of this protected species adversely affected in the past by water pollution. Today it is mostly hard winters that have the greatest impact on Kingfisher numbers, though they can rapidly bounce back in favourable years given their ability to raise two to three broods during the breeding season.

## Added interest

- Kingfishers stun small fish by whacking them against a perch, then swallow them headfirst to avoid fins and dorsal spines lodging in their throats.
- Chicks are raised in a nest cavity at the end of a metre-long tunnel excavated in a bank and feed in rotation, moving to the back of the queue when they have a full stomach.
- Once fledged, young Kingfishers are rapidly driven out of the parents' territory (some before they have learnt to fish) and only half will survive more than a week or two.

## Not quite a Kingfisher?

Britain is not exactly spoilt for choice when it comes to small, vivid blue river birds, so the Kingfisher is unlikely to be confused with anything else. Perhaps our closest approximation is the tree-dwelling Nuthatch, with its blue-grey back, rufous front and dagger bill. However, these tree dwellers lack the dazzling colours of a Kingfisher and would never be seen plunge-diving in a river (*see Nuthatch*). If you spot a small, dark-bodied, white-chested bird slip under the surface of a stream this is the Dipper – it shares the Kingfisher's aquatic tendencies, but none of its plumage colour (*see Dipper*).

Nuthatch

# Dipper

## Key identification features

- The Dipper is a dumpy little bird of fast flowing rocky streams and rivers.
- Shorter and plumper than a Thrush, it has blackish-brown plumage, a conspicuous white chest and, as the name suggests, habitually bobs up and down.
- In flight, it whizzes low along waterways on short rounded wings, often making a sharp high-pitched call and is famed for its ability to hunt invertebrates underwater.

## Where could you find one?

Dippers don't just like water, they like sparkling water – clear, rushing streams and swift-flowing rivers, rich in oxygen and insect life. Found in the north and west, they are lively birds of tumbling torrents that wind between trees and over bare rocks, nesting in rock crevices or overhangs and under stone bridges. During freezing winters some move downstream to lowland stretches of water.

**Juvenile**

## Is it easy to identify?

A short, tubby, bobbing black-and-white bird in the middle of a stream, that's the Dipper. It looks slightly like a photo negative impression of a stocky monochrome Robin. The white bib covering the throat and chest is striking, while the body plumage is largely black on the back, wings and tail and brown on the head and belly. Some riverside views may amount to little more than a dark, short-tailed bird dashing past on whirring wings, uttering a piercing call as it goes. At other times Dippers can prove obliging, perching on a mossy boulder mid-stream and gently dipping up and down, unfussed about being watched. They spend much of the day foraging along their river territories and if you are patient you should be able to watch one nudging down into the current, disappearing from view as it searches for invertebrates beneath the surface.

## How likely am I to see one?

At home amid tumbling water

Dippers are not uncommon in suitable upland river habitats, with around 12,000 pairs in western and northern areas of Britain. They seldom stray far from streams and shallow rocky rivers and can be spotted on riverbank strolls. Along popular walking routes Dippers may be accustomed to passers-by, offering decent views.

## What makes it special?

For ducks, divers, gulls and waders, getting wet comes with the territory, while perching birds, such as thrushes, finches and warblers keep their feet firmly on dry land. Not so the Dipper. It has torn up the rule book, left its chirping relatives on the bank and plunged in. It is one of our most fascinating and unique species: a songbird that swims. Feeding on water insects and their larvae among the rapids means fully submerging in order to search the stony riverbed. Dippers either dive or wade head-down into the water, gripping onto the bottom with strong claws while tilting their wings against the flow to stay submerged. They can remain underwater for half a minute, emerging from the foaming cascade onto a rock, or even swimming at the surface to the shallows – a surreal sight. Dense plumage insulates them against the cold, while muscular legs and wings enable Dippers to hold their position in currents that would sweep similar-sized birds away. A remarkable bird and a wonder to watch.

## Added interest

- The Dipper has a large preen gland at the base of the tail which produces an oily secretion used to prevent its feathers becoming waterlogged.
- Dippers not only possess a clear membrane that can be drawn across the eye for added protection, but also have bright white eyelids which are noticeable whenever they blink and are possibly used as a form of communication.
- Pairs build dome-shaped mossy nests close to the water, with some even raising chicks in crevices tucked away behind waterfalls.

## Not quite a Dipper?

If your 'Dipper' is larger and slimmer than expected and is away from the waterside, then it could be a scarcer upland species called a Ring Ouzel. This summer visitor, numbering a few thousand pairs, looks just like a Blackbird with a white crescent on the chest (*see Blackbird*). Interestingly, Dippers were also once known as Water Ouzels.

The Common Sandpiper might also cause confusion, given it lives in similar areas and shares the Dipper's habit of bobbing up and down. It is a smallish wader, with a grey-brown back, white underparts and a quite long straight bill, and it flies low on stiff, flickering wings marked with a white streak. Around 15,000 pairs breed in north and west Britain.

Common Sandpiper

# Grey Wagtail

Female

## Key identification features

■ The Grey Wagtail is a slender and attractive long-tailed bird, with noticeable bright yellow feathering under the tail.

■ The male has a grey back, a lemon-yellow front and a smart black throat, while females and young birds have less yellow underneath and lack a black throat.

■ Grey Wagtails have a very long black tail, which they habitually wag up and down. In flight a yellowish rump is visible as well as white outer tail feathers and a thin white streak along the centre of the wing.

## Where could you find one?

Nothing is more appealing to breeding Grey Wagtails than fast-flowing fresh water. In the spring and summer they nest alongside rocky upland streams and rivers, especially those bordered by trees, while during the winter they seek out lowland stretches of water of all kinds and will visit canals, flooded gravel pits, lakes and coastal marshes. A widespread resident, the Grey Wagtail tends to be found in northern and western uplands through the breeding season, moving downhill and south during the colder months.

## Is it easy to identify?

The name doesn't really do justice to this brightly-coloured bird with its vibrant lemon zest plumage. However, the 'yellow' title has been claimed by its close relative the Yellow Wagtail, so it has been left with the rather drab sounding 'grey' label, describing its slate-coloured back. It is a common mistake to see a wagtail with yellow plumage and jump to the conclusion that it must be a Yellow Wagtail, when it is actually more likely to be the widely distributed and commoner Grey Wagtail. Telling the two species apart can be tricky and may depend on when and where the bird is spotted. The scarcer Yellow Wagtail is the yellowest of the pair, sporting golden-green upperparts and a buttercup front and face (*see right*). The Grey Wagtail has pinkish-brown legs, rather than black and by far the longer tail, which it almost constantly bobs up and down.

## What makes it special?

In keeping with the name, our trio of wagtails – grey, yellow and pied – all repeatedly twitch their tails up and down and the longest-tailed of the bunch, the Grey Wagtail, is visibly the waggiest of them all. But why expend so much energy beating out baton rhythms with your back end all day long? It has been suggested that a waving tail flushes waterside insect prey out of cover into view, helps flag up an individual's presence to other wagtails along the noisy corridors of rushing streams or signals it is fit and alert to any watching predator. Either way, the Grey Wagtail puts such effort into its pumping action that if you grabbed one by the tail you could imagine the whole body continuing to wag up and down. This lively bird adds movement and colour to river settings and its fidgety nature seems very much in keeping with the species' home among restless river rapids.



### How likely am I to see one?

The Grey Wagtail is widespread and fairly common year-round, numbering 35,000 pairs in Britain. It is easiest to see on a walk along upland watercourses during the spring and summer. During the winter they turn up near water pretty much anywhere and may be fairly approachable, and sometimes visit gardens.

## Added interest

- Grey Wagtails mainly eat insects found along waterways, such as mayflies and beetles, but even tackle tiny fish, small snails, freshwater shrimps and tadpoles.
- Stone bridges are a popular place for Grey Wagtails to nest, as they seek crevices close to running water and are not too fussy whether they are natural or manmade.
- The range of the Grey Wagtail has expanded over the last 150 years from former upland strongholds in northern Britain, yet numbers have fallen in recent decades and it is considered a bird of conservation concern.

### Not quite a Grey Wagtail?

If your wagtail has an olive back, black legs and yellow on the face as well as the front and belly then it could be a Yellow Wagtail – a summer migrant found in meadows, damp pasture and arable farmland in central and eastern England (*see Is it easy to identify? left*). On the other hand, if it lacks yellow and is grey, black and white then it is probably a Pied Wagtail, which is resident in Britain all year round (*see Pied Wagtail*). The Pied Wagtail is the most abundant of our wagtails, with hundreds of thousands of breeding pairs in Britain, while the Grey Wagtail numbers in the tens of thousands of pairs and the Yellow Wagtail in the thousands.

**Yellow Wagtail**

# Great Crested Grebe

Breeding plumage

Winter plumage

## Key identification features

- The Great Crested Grebe is a slender and striking bird, known for its elaborate courtship displays.
- It sits fairly low in the water, dives often and is equipped for fishing with a long neck and dagger-like bill. In the spring and summer, adults of both sexes have a dark-fringed chestnut ruff on either side of the head and a crown of black plumes. The front is white and the dark back fades to brown at the water line.
- In the winter they lose the head trimmings and become grey and white, with the back of the flat black crown showing a hint of breeding plumage tufts.

## Where could you find one?

During the spring and summer, Great Crested Grebes breed on fish-rich freshwater lakes, flooded gravel pits, slow-moving rivers and reservoirs with shallow and weedy areas – mainly in central and eastern England and southern Scotland. In the winter they can also be found at the coast in sheltered inshore areas.

## Is it easy to identify?

There are few waterbirds quite as striking as the Great Crested Grebe in its finest breeding plumage. Even at a distance they stand out from the crowd, with their ornate head feathering, white chest and un-duck-like profile. Their tailless body breaks the surface in a smooth curve and the neck is long and sinuous. The head tapers like a spearhead, from the wide orangey-brown sideburns and backward-sweeping black crown to the sharp point of the long bill. When the head ruff and crest are raised atop a straight slim neck, the faintly glove-puppet shape is unmistakable. Legs and feet are set far back on the body to aid propulsion underwater and in flight they trail behind while the head and neck stick out in front, with panels of white visible on the wings. In the winter, all colour drains from the plumage and the cheek frills disappear.

## What makes it special?

Few British birds put on quite such a show when it comes to courtship. Not only do breeding Great Crested Grebes look extravagant, but pairs engage in complex and elaborate displays on the water. Like synchronised dance partners, the male and female perform a series of choreographed moves, including facing one another shaking their heads vigorously, or submerging before rising, chest to chest with beaks full of water weed while paddling furiously to remain vertical.

Unfortunately the decorative head feathering that sets them apart also attracted the attention of Victorian fashion designers and almost led to their extinction in Britain. Their plight, along with that of other wild species harvested for the fashion industry, led to the passing of bird protection laws and the founding of the RSPB. They have since bounced back, taking advantage of the flooding of gravel pits and creation of new reservoirs during the twentieth century and now grace open water across much of Britain.

### How likely am I to see one?

More than 4,500 pairs of Great Crested Grebes breed on lakes in Britain and can be relatively easy to view, with courtship displays reaching their peak between late winter and spring. Numbers double in the colder months as birds from Europe head here to avoid winter conditions further north.

## Added interest

- Great Crested Grebes lay their eggs on an anchored floating raft of water vegetation, covering them up whenever they are away from the nest to hide the clutch from predators.
- They tend to dive when threatened rather than take flight and need a lengthy run-up in order to get airborne, so are found on large lakes rather than small ponds.
- The young are marked with black and white stripes like humbugs and hitch a ride on their parents' backs for warmth and protection.

### Not quite a Great Crested Grebe?

The four other species of grebe in Britain are all smaller than the Great Crested Grebe and look very different in breeding plumage. Three of the grebes are scarce, while the most common of the bunch is a dumpy little bird called the Little Grebe, also known as a Dabchick (*see Coot*). Confusion is more likely with various larger aquatic birds that sit fairly low in the water and share the same long-necked shape and pointed bill as the Great Crested Grebe. They include 'sawbill' ducks such as the female Goosander and Red-breasted Merganser, which have tufty brown heads, and 'divers' such as the Red-throated Diver, which looks very similar in dark grey and white winter plumage when thousands are around our coast.

**Red-throated Diver in winter plumage**

# Moorhen

Juvenile

## Key identification features

- The Moorhen is a perky, pigeon-sized waterbird with dark glossy plumage.
- It has a red beak tipped with yellow and the base of the bill extends in a 'shield' up the forehead. The body feathering appears blackish, but in good light a subtle brown tone to the back and inky blue hue on the head, neck and front is evident, separated by a broken white line along the side.
- The tail is white underneath with a black central band and the legs and long toes are greenish-yellow.

## Where could you find one?

Almost any stretch of still or slow-moving water with well vegetated banks is suitable for Moorhens, whether in rural or urban settings. They can be found breeding on lakes, rivers, reservoirs, canals, marshes and park ponds and are only scarce or absent in uplands and northern Scotland. They feed on plants, seeds, berries and invertebrates around the edges of open water and in neighbouring fields.

## Is it easy to identify?

Moorhens have quite distinctive jerky movements as they go about their business, walking with high steps to avoid tripping over their long greeny-yellow toes and nodding their heads as they swim. They twitch their tail upwards, the underside a flash of white split by a vertical black stripe. Most noticeable of all is the bright yellow-tipped glossy red beak, which helps distinguish the Moorhen from the closely-related Coot. This slightly larger, slate-black waterbird has a white beak and a white forehead 'shield', which gave rise to the term 'bald as a Coot'. Coots prefer deeper water, diving frequently to feed and can be seen in the centre of lakes, while Moorhens skirt around the fringes. Coots also have odd-looking grey toes that have leaf-like lobes to aid propulsion, while the Moorhen's yellowish toes are long and spindly. Both Moorhens and Coots frequently engage in aggressive territorial boundary disputes.

## What makes it special?

Attractive and adaptable, Moorhens can be found in and around almost any vegetated stretch of water and have overcome natural wariness to take advantage of suitable habitat in urban areas. As such, they are one of our most successful freshwater species and add a sense of novelty among the ducks and geese on a city lake. Different enough to demand identification, they provide the perfect introduction to birdwatching, being generally easy to observe whether feeding, nesting or with fluffy chicks in tow and satisfying to distinguish from the similar-looking and widespread Coot (red beak for Moorhen, white for Coot). Moorhens are strong characters, given to bickering and brawling with one another in noisy territorial skirmishes. Improbable as it sounds, these long-toed waterbirds are also agile climbers, capable of nesting in trees and bushes.

### How likely am I to see one?

The Moorhen is a widely distributed and familiar resident, numbering around 260,000 pairs in spring and summer. The greatest concentrations are found in central and eastern England and the population swells during the colder months as birds from mainland Europe fly here to escape freezing winter weather.

## Added interest

- Moorhens are not associated with barren upland moors – their name derives from the related term 'mere' meaning a lake or marsh.
- Pairs raise two or three broods a year and older offspring may help with the feeding of newly-hatched chicks.
- Although Moorhens can fly to escape danger, they also dive, briefly remaining hidden underwater by gripping onto vegetation or the bottom in order to overcome their natural buoyancy.

### Not quite a Moorhen?

If your Moorhen looks dull and brown, then it may well be a juvenile that has yet to gain the handsome plumage. If its beak and forehead are white rather than red, then you are looking at a resident Coot (*see Coot*). There are two other possibilities that share the same habitat. The Little Grebe is a dumpy dark-coloured waterbird that has a pale fluffy rear end, black cap and chestnut cheeks. It numbers around 5,000 pairs in summer and 15,000 birds in winter. The scarcer Water Rail is a slender, skulking bird of waterside vegetation with a streaky brown back, grey head and underparts and long red bill.

Water Rail

# Coot

## Key identification features

- The Coot is a black waterbird, with a plump, rounded shape and a conspicuous white beak and forehead.
- Both males and females look the same, but juvenile birds are paler and have a light face and neck.
- Coots spend most of their time on the water, nodding their heads as they go and paddling with greyish feet that have odd-looking lobed toes.

## Where could you find one?

A widespread bird of open freshwater habitats, the Coot can be found on reservoirs, flooded gravel pits, slow-moving waterways and lakes in rural and urban areas, though is absent from uplands and the far north-west. In the winter birds gather in flocks on larger water bodies, which are less likely to freeze over, and sometimes visit estuaries and harbours.

## Is it easy to identify?

The simplicity of its appearance makes the Coot a straightforward bird to identify, being matt black all over with a white beak. Good light reveals the body plumage to be pencil-lead grey in tone, while the white of the pointed bill extends up the forehead in a frontal 'shield' that stands out even at a distance. Although it swims around much like a duck, often diving to feed, the Coot is instead a member of the rail family, which includes the similar Moorhen. This slightly smaller species can be told apart from the Coot at a glance by its red beak and flash of white on the tail underside (*see Moorhen*). Their toes are also different: thin and spindly for the Moorhen, which is more adept at getting about on land, and large with leaf-like lobes for the Coot, a strong swimmer in deep water.

## What makes it special?

A lake on a calm day can seem a tranquil place – until the Coots start bickering. Quarrelsome and territorial by nature, these black and rotund birds, like cannonballs with short fuses, engage in explosive boundary disputes, noisily charging at neighbours and grappling violently with their feet. There's nothing cute about a Coot in a foul mood. And they can even turn their aggression on their own young, occasionally killing weaker members of the brood. Outside the breeding season they are much more gregarious and gather on open water in flocks that can be thousands strong, making for an impressive sight. Confident characters, with uniform dark plumage and diagnostic white on the forehead resembling a splodge of sugar icing, they always look smartly turned out. They can also lay claim to having the strangest feet of any of our birds!

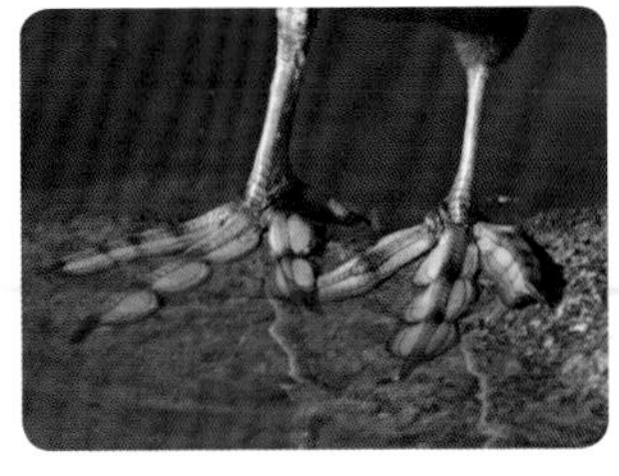



### How likely am I to see one?

The Coot is commonly encountered on lowland water bodies, with the greatest concentrations in central, southern and eastern England. It has readily colonised newly-created reservoirs, wetland nature reserves and flooded gravel pits, and currently numbers 30,000 pairs during the breeding season, with visitors from northern Europe boosting the winter population to 180,000 birds.

## Added interest

- The species' prominent white forehead is believed to have given rise to the centuries-old expression 'bald as a Coot'.
- Although reluctant to take to the air, and requiring a running take off, they can migrate a fair distance – some even crossing the Channel to spend the winter in Europe.
- Chicks soon leave the nest after hatching and parents may split the brood between them as they feed the young for about a month.

### Not quite a Coot?

The Coot is most likely to be confused with the Moorhen, which has a red beak and forehead (see *Moorhen*). Young birds of both species that have yet to develop their distinctive bill colour are very similar, although the juvenile Coot is much paler on the throat and exhibits the species' characteristic tailless shape. This blunt-ended appearance is shared by another smaller waterbird, the Little Grebe, also known as the Dabchick, which is widely distributed with a population of 10-15,000 birds. A shy brown species that regularly dives, it has chestnut cheeks in the breeding season, a pale powder-puff rear end and makes a whinnying sound like a horse.

**Moorhen**

# Grey Heron

## Key identification features

- The Grey Heron is a stately grey, black and white bird with long legs, neck and beak.
- It has a formidable pointed bill and a badger-stripe of black above each eye that tapers behind in a thin and flimsy crest. The body is pale grey, with broken black streaks down the front of the neck and light wispy feathering on the chest and back.
- In flight the legs trail behind, the head and neck are pulled in and the broad, bowed grey wings (spanning over 6ft) are dark across the ends and trailing edges. While the sexes are alike, juveniles are more uniformly grey.

## Where could you find one?

Freshwater habitat of all kinds will attract herons, ranging from fish-rich wetland marshes and lowland lakesides to slow-flowing rivers and well-vegetated reservoirs. These widespread residents can also be found around the coast, hunting on estuaries or among seaweed-covered rockpools and will even visit gardens to raid ponds stocked with goldfish.

## Is it easy to identify?

The Heron towers above other birds as it paces the fringes of a lake or marsh looking for prey. It can be overlooked when standing motionless at the water's edge waiting patiently for fish to pass within striking range, but as soon as it moves it is unmistakable, with its long legs, sinuous neck and dagger bill. It is built for stabbing at fish with lightning fast lunges – eyes set well forward and head tapering to a point in the fearsome blade of a bill. The dark stripe over each eye narrows behind in fine strands, resembling the trailing threads of a black bandana. In flight its broad wings are distinctive, it trails its legs behind and tucks its head in close to its body. The loud call is a short, harsh bark that has been described as sounding like someone shouting *Frank*!

## How likely am I to see one?

The Grey Heron benefitted from the cleaning up of polluted waterways during the twentieth century and is today a widespread and relatively common bird, with around 12,000 pairs in Britain. Our Herons are present year-round, like to hunt alone and nest together, and can be encountered on almost any suitable stretch of water or at noisy treetop breeding colonies.

## What makes it special?

The Heron is a supreme angler, combining limitless patience, extreme stealth and breathtaking speed. Its favoured technique is simply to pick a spot, keep still and wait until unsuspecting fish swim by. Front on and viewed from below, the slender neck leaning over the water's surface might easily be mistaken for a stick, with little to signal danger but a pair of yellow eyes peering down the barrel of the beak. With a sudden thrust it strikes, grasping slippery prey which is typically swallowed whole and headfirst. It may also take a more active approach to finding a meal, stalking fish and amphibians in the shallows or wading deeper to flush them into view – items on the menu even include ducklings, crabs and, on land, weasels, rats and moles!

## Added interest

- During the Grey Heron's breeding season, which may start as early as February, the adult's dull yellow bill becomes flushed with colourful pinkish-orange hues.
- Herons were once prized and protected by nobles as challenging quarry for falconry, given their size and surprising agility in the air, while anglers would rub heron oil on baits and lines hoping it would improve their chances of catching fish.
- A staggering 400 roast Herons were among thousands of dishes served up at an extravagant medieval banquet laid on at Cawood Castle by George Neville on his appointment as Archbishop of York in 1465.

## Not quite a Grey Heron?

If your Heron is smaller than expected and pure white, then it is an Egret. Little Egrets pushed north from mainland Europe into Britain and began breeding by the mid-1990s, gradually increasing in range and population – winter migrants boost numbers into the thousands. A black bill and black legs contrast with the snowy white plumage and they have yellow feet. If, on the other hand, your heron is light brown rather than grey, you may have struck lucky and chanced across a Bittern. An excellent sighting! This shy and well-camouflaged streaky reedbed inhabitant numbers several hundred individuals in Britain and is most easily encountered at wetland reserves in winter, when freezing weather forces them out from cover to find fish in more exposed water.

Little Egret

# Waders

# Oystercatcher

## Key identification features

- The Oystercatcher is an eye-catching black-and-white wader, with bold markings and a loud piping call.
- Males and females look alike and are black on the head, neck, breast and black and white underneath. They have a long, straight orangey-red beak, red eyes and pinkish legs.
- In flight the wings are black on top, with a prominent broad white stripe, while the white rump runs up the lower back in a pointed triangle. During the winter, Oystercatchers have a white half-collar beneath the chin.

## Where could you find one?

Oystercatchers can be seen throughout the year all around our coast. They nest on grassy, sandy or pebbly shores and forage across tidal mudflats and around rocky bays for shellfish, marine worms and other invertebrates. Some pairs also breed inland close to rivers and lakes, particularly in Scotland and northern England, but move to the coast in winter, where large congregations gather to roost.

## Is it easy to identify?

The Oystercatcher is not exactly a discreet bird. In fact, it does its level best to be noticed. If it doesn't catch the eye striding around in its striking black-and-white plumage, candy-orange beak sticking out in front, then its shrill far-carrying *kleep, kleep* call is sure to grab the attention. Few of our birds are quite as conspicuous, which makes identification straightforward even at a distance or in flight. And Oystercatchers keep things simple by looking pretty much the same whichever sex and whatever time of year – the only changeable feature being a white chinstrap (like a wide vicar's collar) that adults sport in early winter and which is also present in young birds. While not a tame species, Oystercatchers are fairly confident and tolerant of people, enabling good views to be had.

## How likely am I to see one?

A parent bird with chicks

An estimated 110,000 pairs breed around Britain and during the colder months our population is swelled by over-wintering European birds from the north and east. Oystercatchers are common and widespread along our coastlines and, in the spring and summer months, nesting pairs can also be seen inland in northern Britain.

## What makes it special?

These chunky, characterful waders stand out in a crowd, piercing the air with their sharp peeping calls. That brightly-coloured beak is not just for show, but is a powerful tool which allows them to target a range of different prey. Feeding along tidal stretches, the Oystercatcher dips its long bill into the muddy sand, like a carrot stick into hummus, in search of marine worms and molluscs and also forages amid seaweedy rocks for shellfish and crustaceans. Their main food is cockles and mussels, not oysters as the name suggests, and they either pierce then prise open the tightly-clamped shells or break them apart with hammer blows. As chicks are unable to get into the tough bivalves, they rely on their parents for food, at the same time learning the adults' favoured techniques for tackling shellfish: precision stabbing or hammering. The result is that the different approaches are passed on from one generation to the next, with specialist hammerers gradually developing blunter bill tips as a result of all that pounding. Whatever method they employ, Oystercatchers typically live for over a decade, with records of individuals surviving more than 40 years.

## Added interest

- Territorial pairs and small parties of Oystercatchers engage in piping displays, bowing their heads with open bills pointing to the ground and calling loudly in a frenzied ear-splitting chorus.
- Numerous factors cause shellfish populations to fluctuate. The fact that an Oystercatcher can eat several hundred cockles a day has in the past led to culls in commercially-exploited bays.
- While pairs generally nest on the ground, they have also been recorded raising chicks on gravel-covered flat roofs and roundabouts.

## Not quite an Oystercatcher?

The Oystercatcher's combination of pied plumage, a stocky build and orangey-red bill make this a fairly unmistakable species. However, it is not our only black-and-white wader. The long-legged and elegant Avocet, with black over the top of the head and around the folded wings, is an exciting sighting and has a uniquely-shaped dark bill that curves upwards at the end. Once extinct as a British breeding species, it is still scarce and mainly nests by coastal lagoons in eastern England. However, numbers rise above 7,000 during the winter, when gatherings concentrate at estuaries in the south and west. Another wader that shares the Oystercatcher's white stripe across the open wings and has a reddish-tinged bill, is the Black-tailed Godwit (*see Black-tailed Godwit*).

Avocet

# Curlew

## Key identification features

- The Curlew is a large and distinctive wading bird, with a seagull-sized body, lanky grey legs and a very long bill which is smoothly down-curved like half a longbow.
- The plumage is mottled brown on the back, streaky on the front and pale underneath. In flight the open wings are slightly darker towards the ends, while the white rump extends up the back in a pointed triangle.
- It has a distinctive bubbling song and a melancholic call.

## Where could you find one?

Resident pairs nest on upland moorland, rough pasture and marshland areas, mainly across the north of Britain, though also in scattered southerly areas. In the autumn and winter, Curlews gather all around our coast, feeding along the shoreline or in fields nearby, with the largest numbers found on major estuaries.

## Is it easy to identify?

The Curlew stands out from the crowd not just because of its size, but also its preposterously elongated bill. Take away the beak and this is a bulky bird with straightforward plumage. The spangled and speckled brown feathering may look pale and monochrome or warm and golden, while its long legs are the colour of the grey estuarine ooze it strides through in search of marine worms and crustaceans. When it takes flight, the Curlew's triangular white rump running up the back, pointing the way forward, looks as if a seam has split in its brown outfit, while the pale underside of the wings and belly contrast with its streaky straw-coloured front. It also utters a *coor-lii* call, from which it gets its name and, during the breeding season, delivers a series of notes that accelerate in a trembling crescendo.

## How likely am I to see one?

Chicks hatch with short straight bills

Over 65,000 pairs nest in Britain, with most breeding birds concentrated in boggy and grassy upland areas of Scotland and northern England. In the autumn and winter, our resident population heads for the seaside and is boosted by visiting Curlews from Scandinavia and northern Europe. Curlews can be wary birds and not easy to get close to, so binoculars are recommended for good views.

## What makes it special?

On paper there is something faintly ludicrous about the Curlew's appearance. The disproportionately long drooping beak looks as though it has lost a battle with gravity, like a bill-stretching experiment that went wrong before hardening. Yet out in the wild it makes perfect sense – allowing the Curlew to probe deeper for worms than other species. It is also a dextrous tool that enables this leggy wader to winkle out shellfish from crevices, pick up insects and even dismember troublesome crabs at a safe distance without risking an eye injury. The Curlew has a sense of wilderness about it and its song is one of the most evocative in the bird world: a series of simple escalating notes that come to a rapid boil in a bubbling trill. For some, it is joyous and uplifting, while for others the Curlew's quavering cry is plaintive and sad – the soundtrack to spring on desolate upland moors where it nests. Either way, this is a bird well worth hearing as well as seeing.

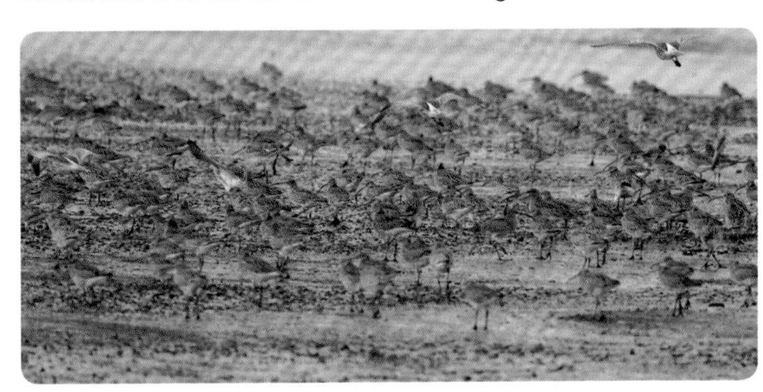

## Added interest

- While adult Curlews have much the same plumage whatever the sex or time of year, the female is larger and heavier than the male and also has a longer beak.
- A drastic population decline over recent years means the Curlew is now considered a species of high conservation concern in Britain, which hosts internationally-significant numbers during the winter.
- America's once-abundant Eskimo Curlew was driven to extinction by the middle of the last century through over-hunting and habitat loss, while Siberia's Slender-billed Curlew has not been seen in years and may have headed the same way.

## Not quite a Curlew?

A similar-looking, but much rarer, relative of the Curlew is the Whimbrel. This exciting find turns up along the coast in scattered locations. It is best identified by subtle dark and light stripes on its head and its rapid-fire whistling call. The Whimbrel's beak also bends downwards towards the end, rather than in a more gradual curve. If your Curlew looks like someone has been having a go at its beak with the straightening tongs, then it could be a kind of godwit. A tad smaller than the Curlew, but big all the same, the leggiest, with white wing bars and a black-ended white tail, is the Black-tailed Godwit (*see Black-tailed Godwit*). The other, with plainer wings and shorter legs, is the Bar-tailed Godwit.

Whimbrel

# Black-tailed Godwit

Breeding plumage

## Key identification features

- The Black-tailed Godwit is a large and elegant wader, with a lengthy straight beak and long legs and neck.
- In breeding plumage, adult birds are orangey-brown across the head, neck and chest, and the pale belly is barred across the flanks. During the colder months, when they are most numerous, they lose the rufous colouring and look plain grey-brown.
- In flight, Black-tailed Godwits have a noticeable white stripe along the centre of the dark upper-wing, and a broad black band runs across the end of the white tail.

## Where could you find one?

Only a few dozen pairs breed in Britain during the spring and summer at scattered damp and grassy locations, mainly in East Anglia. In the autumn and winter, they are far more common, congregating at muddy and marshy habitats around Britain, such as coastal lagoons, estuaries and wetlands, in particular in north-west England and the south and south-east of the country.

## Is it easy to identify?

The lanky Black-tailed Godwit stands out from the crowd on account of its size. One of our largest waders, it can be told apart from similar-scale species, such as the Curlew, by the shape of its bill. Straight and elongated, this black-tipped tool is ideally suited to probing wet soil and silt for invertebrates, and they happily wade deep into water to feed. The Black-tailed Godwit is a slender and attractive bird – especially so in the breeding season when the head and front are flushed rusty-orange. They are mostly seen during the colder months in dowdier greyish-brown plumage, at a time of year when another similar Godwit is also present on coastal mudflats: the Bar-tailed Godwit (*see right*). Whatever time of year, the Black-tailed Godwit has a striking white stripe along its dark wings, a white rump and conspicuous black end to the tail.

## How likely am I to see one?

Distinctive markings in flight

This is a relatively rare breeding bird in Britain, but fairly common and widespread outside the nesting season when large numbers from Iceland visit our shores. More than 40,000 spend the colder months on our frost-free lagoons and estuaries, which can attract large flocks, while thousands more stop over briefly on migration to and from wintering grounds further south.

## What makes it special?

Every pair of Black-tailed Godwits that raises young in Britain is a small cause for celebration. This handsome wader was driven to extinction as a breeding bird by the mid-1800s as a result of overhunting and the drainage of wetlands, and it took a century before they made a successful comeback after a pair nested in the Ouse Washes, East Anglia, in 1952. Since then this protected species has gradually become re-established, particularly on nature reserves, and the number of birds spending the winter here has risen. Black-tailed Godwit pairs can stick together over a lifetime - although that depends on their punctuality when it comes to keeping a date. Males and females may pass the winter months far apart but arrive back on their nesting grounds at almost exactly the same time. Research has discovered that those dawdlers who return late for the reunion face being 'divorced' as their impatient partner seeks a new mate.

## Added interest

- Females are slightly larger than males and have a longer beak, which may help prevent pairs from directly competing with one another for food.
- Two distinct races of Black-tailed Godwits occur in the UK: European birds that nest here and winter in Iberia and Africa, and visiting Icelandic migrants, which are more richly coloured in the breeding season.
- Godwits hold the record for the longest non-stop migration flights of any bird, with one tagged female Bar-tailed Godwit covering 7,200 miles between Alaska and New Zealand in eight days.

## Not quite a Black-tailed Godwit?

Outside the breeding season our Black-tailed Godwits are joined around the coast by similar-looking Bar-tailed Godwits. Both birds have grey-brown winter plumage and share the long straight beak (though the Bar-tailed Godwit's is ever so slightly upturned). The clue to separating these lookalike relatives is in their name: black tail tip for one, lightly barred for the other. In flight the Bar-tailed Godwit lacks a white wing stripe and on the ground it stands shorter, showing less leg above the 'knee'. If your Black-tailed Godwit has a down-curved beak then you are looking at a Curlew or Whimbrel (*see Curlew*), and if it is on the small side and has red legs then it could be a Redshank (*see Redshank*).

Bar-tailed Godwit

# Redshank

Winter plumage

##  Key identification features

- The Redshank is a medium-sized brownish wader, bigger than a Blackbird, with gangly orange-red legs and a straight, fairly long bill, which is red at the base and dark at the end.
- During the spring and summer breeding season, both sexes have a mottled brown back, a streaky front and whitish underparts, but are plainer and greyer in the autumn and winter. They also have a thin pale ring around the eye.
- In flight the wings have a white trailing edge and the white rump tapers up the back to a point.

##  Where could you find one?

This is a widespread and relatively common coastal species, which can also be encountered in wet and boggy places inland. During the breeding season the greatest concentrations are in Scotland and northern England. Pairs nest mainly on saltmarshes, though also away from the coast on rough grassland, damp meadows and wetland reserves. In the autumn and winter Redshanks are found around the coast, on mudflats and shores.

##  Is it easy to identify?

As the centuries-old name suggests it has red shanks: scarlet liquorice-sticks that stand out particularly well as it forages in a busy manner on marshes and tidal stretches. Its beak is also noticeably red, though not for the entire length, as if darkened by probing in estuary mud. Take away the coloured legs and bill and the body plumage is not particularly attention-grabbing. Darker on top and lighter beneath, it is either brown with a streaky front in summer, or a more uniform smudgy-grey in winter. This wary and vocal species gives itself away in flight whatever the time of year. Not only does it have a pointed white rump, but also a distinctive broad patch of white on the rear edge of the wings. On landing, it can pause momentarily with wings stretched aloft, flashing the white undersides.

## How likely am I to see one?

Revealing white underwings in flight

Roughly 25,000 pairs of Redshanks breed in Britain, mainly in the northern half. While some individuals, especially young birds, head south to the Continent during the colder months, our remaining residents are joined by over-wintering visitors from Iceland, Scandinavia and northern Europe, with numbers building to around 160,000 birds nationwide by mid-winter.

## What makes it special?

The Redshank is an attractive and elegantly-proportioned bird, whose orange-red legs add a dash of colour on a grey tidal mudflat as it feeds briskly on a variety of small invertebrates. It is also a highly-strung species, quick to take flight, uttering sharp calls at the merest hint of danger and as a result has earned the nickname 'sentinel of the marshes'. Other birds in the company of Redshanks can relax in the knowledge this anxious early warning system is fretting on their behalf, going off like an over-sensitive car alarm at any approaching threat. In the breeding season, Redshanks may even perch on a vantage point such as a wall or fence post, twitching and bobbing as they watch for intruders. Panicky nature aside, the Redshank has a lovely piping *teu-u-u* call, a familiar and atmospheric sound of quiet coastal bays and exposed marshes.

## Added interest

- Redshanks feed on a host of small invertebrates, but are particularly fond of the tiny bite-sized Corophium mud shrimp that can live in dense concentrations on tidal estuaries, numbering thousands per square metre.
- In the winter, Redshanks remain faithful to their favourite coastal feeding sites, whatever the weather, which can lead to high mortality among young inexperienced birds during freezing spells.
- Although they typically live for around four years, the oldest Redshank recorded in Britain was ringed as an adult in Kent in 1987 and lived for a further 20 years before being killed by a cat on South Uist, in the Western Isles, in 2007.

## Not quite a Redshank?

Other medium-sized wading birds with longish straight bills and coloured legs, which could be confused with the Redshank, include the scarcer Spotted Redshank, which lacks white wing bars, the Ruff and also the smaller and shorter-billed Turnstone and the Purple Sandpiper (*see Turnstone*). If your Redshank has greenish legs, then it could be the aptly-named Greenshank. Up to 1,500 pairs nest in boggy areas of northern Scotland, though outside the breeding season they can be seen beside freshwater and the coast all around Britain, with many passing through on migration. They are pale and elegant waders, sporting a greyish back, white underparts and a long wedge of white rump up the back that is visible in flight.

Greenshank

# Ringed Plover

## Key identification features

- The Ringed Plover is a small, boldly marked wader with a black-and-white head and orange legs. It has a grey-brown back and is white underneath.
- The white neck is encircled with a black band, which broadens over the breast and a black mask runs across the eyes and face. There is a horizontal white patch on the forehead, just above the base of the short, black-tipped orange bill and in flight a light stripe is visible along the brown upper-side of the wings.
- Both sexes look much the same and in the winter their crisp black head markings become dull and browner, much like those of juveniles.

## Where could you find one?

Ringed Plovers can be found pretty much all around our coast throughout the year. They lay their eggs on the ground in areas of sand or shingle and some nest inland beside flooded gravel pits and other suitable stretches of water. Outside the breeding season they congregate at the coast, feeding on shorelines and muddy estuaries and are joined by visitors from the north.

## Is it easy to identify?

This attractive little orange-legged species is a relatively common sight on shorelines around Britain and is distinctively 'ringed' with hoops of black and white around the head and neck. Its plumage sets it apart from other waders and it feeds in a tell-tale way, using its large eyes to identify tasty titbits. Instead of motoring along, incessantly probing and pecking as it goes, the Ringed Plover takes a few quick strides before pausing to pick up whatever it has spotted – a stop-start approach akin to a learner driver lurching between accelerator and brake. The Little Ringed Plover is a lookalike relative, however, this slightly slimmer species has a dark bill and lacks thin white wing bars (*see Not quite a Ringed Plover?*). In the winter, any possibility of confusing the twin species disappears as the 'Little' version flies south, leaving us with just our standard Ringed Plover.

## How likely am I to see one?

More than 5,300 pairs breed in Britain, mainly in coastal areas, although also at some inland sites. In the colder months, over-wintering migrants push up numbers to more than 35,000 birds. They are easiest to spot feeding on tidal mudflats and along shorelines, often roosting or flying in small tight-knit flocks uttering a pleasant two-syllable fluty call.

**In flight showing white wing markings**

## What makes it special?

On paper the Ringed Plover looks like it would stand out a mile on a beach, with bands of black and white around its face and neck, yet it provides surprisingly good camouflage when sitting still on a ground nest amid pebbles and shingle. With tangerine legs tucked beneath, the mushroom-brown back blends in and the dark head markings resemble shadows and help break up its outline. Their speckled stone-coloured eggs are virtually impossible to spot, reducing the chance of them being taken by gulls – though at popular coastal spots they run the risk of being trodden on by unwary beach-goers. The chicks must rank among the cutest around: fluffy little salt-and-pepper puffballs tottering around on gangly outsized legs. Ringed Plover parents use deception when a potential predator approaches the nest site, drawing attention and luring the intruder away by shuffling along the ground acting as if they have a broken wing. Once the threat is at a safe distance they fly off, miraculously healed.

## Added interest

- The Ringed Plover is a runner and wader with no need of fully webbed feet like swimming birds, however its two outer toes are very slightly webbed.
- Ringed Plovers from as far away as Canada and Greenland stop off in Britain on migration to and from their wintering grounds in west Africa.
- Beaches popular with day-trippers and dog-walkers are far from ideal places to nest and in parts of England pairs favour coastal nature reserves which offer protection and relative tranquillity.

## Not quite a Ringed Plover?

If your Ringed Plover has the required orange legs, short bill and white underside, but a much darker brown back and messier head and breast markings, then it could be a Turnstone. This similar-sized species actively feeds on shores and seaweed-covered rocks (*see Turnstone*). If it is the spring and summer breeding season and your Ringed Plover is inland by freshwater, has pale legs, a yellow ring around the eyes and lacks a white wing stripe in flight then it is a Little Ringed Plover. The Little Ringed Plover is a dedicated freshwater species of manmade habitats in central and southern England. It first colonised Britain between the wars and this annual visitor spread and increased in number to over 1,200 pairs today.

**Little Ringed Plover**

# Turnstone

Winter plumage

Breeding plumage

##  Key identification features

- The Turnstone is a Blackbird-sized wader, with a tubby body, short orange legs and a dark pointed beak.
- In the spring and summer breeding season, the adults have a white underside, blotchy orangey-chestnut and black markings on the back, a black chest and a patchy black-and-white head. During the colder months they lose the tortoiseshell upperparts and turn dark mottled brown on the back, head and chest.
- In flight they have a pied appearance, with areas of white on the dark back, shoulders, wings and top of the black-tipped tail.

##  Where could you find one?

This is a coastal species that forages on stony beaches, seaweed-covered rocks, muddy shores and along sandy strandlines all around Britain, typically in small flocks. While Turnstones can be spotted in the spring and summer, they don't nest here and the greatest numbers visit from their Arctic breeding grounds during the colder months, either passing through on migration or stopping over for the winter. They are opportunist feeders and can be found around harbours, jetties and quaysides searching for titbits.

##  Is it easy to identify?

In breeding plumage the upperparts are daubed, as if by an abstract artist, with splotches of cinnamon-brown and black, while brushstrokes of white and black pattern the face and chest. This variegated feathering provides surprisingly good camouflage against a backdrop of seaweed and shingle. However, their active nature, chattering calls and orange legs draw attention and they can be remarkably tolerant of people, enabling decent views to be had. By the winter they have lost their brindled back and chestnut tones, becoming a dull dark-brown on top and around the face and breast – though their legs are still bright orange. When they take to the wing in tight weaving flocks, they are dark above, but with a distinctive array of white wedges. Instead of making do with a little white on the tail and open wings, like many small waders, they also have patches on the shoulders and back.

## What makes it special?

If you flip over stones on a beach or kick back strands of seaweed you can be sure to expose a wealth of invertebrate life furiously flipping and scampering about. This is exactly the tactic the aptly-named Turnstone employs to find food. Using its sturdy pointed beak, it flicks over small pebbles, tresses of seaweed, shells and driftwood, then feeds on the hidden life beneath. There are even records of several birds working together to lift larger stones! Turnstones clamber over rocks, search tidal mudflats and comb the strandline for anything they can find. You can often get quite close to these industrious and engaging birds without frightening them off. The vast majority of our Turnstones visit outside the breeding season as they escape freezing conditions on their Arctic nesting grounds and they are remarkable long-distance migrants. In the winter, the Turnstones you are watching fossicking around amid the flotsam and mops of seaweed are likely to have come all the way from Greenland and Canada.

### How likely am I to see one?

Early autumn to early spring is the best time to see Turnstones, with around 50,000 spending the winter on our shores. Birds from northern Europe pass through on migration to Africa, while those from Greenland and Canada stay for the winter. Some non-breeding birds hang around over the summer, but in lower numbers scattered around mainly rocky coastlines in northern Britain.



##  Added interest

- Despite tens of thousands of Turnstones visiting Britain every year, there are no confirmed records of them ever having bred here, despite a couple of reports of possible nesting pairs in northern Scotland.
- Research has discovered that individual variations in plumage patterns around the head and upper body, allow Turnstones to recognise one another by sight.
- The word Turnstone is such an apt description of the species' behaviour that it is also the basis of its name in other languages, from German and Norwegian to Spanish, French, Dutch and Italian.

## Not quite a Turnstone?

If your Turnstone has a lighter, sandy-brown back and much neater black-and-white markings on the head, then it could be a Ringed Plover. This busy little coastal wader also has orange legs and feeds by sight, adopting a distinctive watch-run-peck technique (*see Ringed Plover*). However, there is a similar-sized wader that shares a love of seaweed-covered rocks, often feeding alongside Turnstones: the Purple Sandpiper. This dumpy winter migrant has orange or yellow legs and a longer, slightly down-curved bill, which is yellowish near the base. It has rather drab grey-brown plumage, not particularly purple, a pale belly and streaky flanks. Around 13,000 visit Britain annually, mainly the northern half, feeding along stony shores.

**Purple Sandpiper**

# Dunlin

Winter plumage

Breeding plumage

##  Key identification features

■ The Dunlin is a small, Starling-sized wader, with short dark legs and a longish black bill which curves downwards slightly.

■ It has two differing outfits in its wardrobe. In breeding plumage both male and female Dunlin have a tortoiseshell back of caramel brown and black, and a carpet-tile square of black on the belly. In winter, they are light muddy-grey with a white belly, and during between-season costume changes they can end up with a mix of features as they moult.

■ In flight, a pale line running down the wing is visible and the dark rump has white sides.

##  Where could you find one?

In the winter, Dunlin are common and widespread around our coasts, on mudflats, marshes, estuaries, creeks and sandy shores, where they feed in busy flocks on invertebrates hidden in the mud, sand and seaweed. During the spring and summer breeding season, they nest on boggy upland moors, mainly in northern Scotland.

##  Is it easy to identify?

Waders can be a challenge to identify, but the Dunlin is our most common small brown wader, so if you are inclined to blurt out a name on seeing a bird that roughly fits the bill then 'Dunlin!' is the safest bet. In breeding plumage it is quite handsome: reddish-brown on top and the only wader of its size with a black patch on its belly. However, things get a bit trickier in winter, which is also when they are most abundant. At this time of year Dunlin are fairly nondescript, being uniform grey-brown above and pale below. Their moderately-long bill is an important clue, as it curves downwards slightly. They scarcely look up when feeding, snatching small invertebrates from the estuarine ooze or 'stitching' tidal mud with rapid probes as they walk, and short legs mean they can only wade in very shallow water.

## How likely am I to see one?

**Dunlin can only wade in shallow water**

Winter is the best time to see Dunlin as our coasts attract several hundred thousand, joining other waders in gorging on a daily smorgasbord of shellfish, worms and other invertebrates. Given that they are fairly small and plain-looking, Dunlin can be easy to overlook, especially in the company of more eye-catching shorebirds. During the summer only around 10,000 pairs breed in Britain, the rest departing for more northerly lands.

## What makes it special?

Correctly identifying any small wader is a satisfying step forward for the learner birder, so to confidently recognise a bird as a Dunlin is something to savour. In the winter they gather in flocks, which twist and turn in flight like swirls of smoke over the shoreline, undersides flickering white as they change direction. At high tide they roost out of the water's reach, then spread out to feed on the freshly exposed mud and sand as the sea retreats, their activity dictated by the movement of the tides rather than the passage of day and night. Spectacular flocks can be thousands strong and contain birds which have headed south for the winter from Arctic lands.

## Added interest

- The Dunlin has the nickname 'Plover's page' as it can form close attachments to Golden Plover on upland breeding grounds, benefiting from the Plover's vigilance when it comes to spotting predators.
- Although Dunlin live on average for five years, one bird ringed in Scotland in 1991 was re-caught over 19 years later wintering in Mauritania, west Africa.
- Three races of Dunlin can be found in Britain, which vary slightly in appearance and beak length. Our breeding Dunlin head for west Africa in the winter and are replaced by birds from Scandinavia and Russia, while Dunlin from Greenland pass through on migration, stopping here only to rest and refuel.

## Not quite a Dunlin?

During the colder months things can get confusing, as Dunlin have lost their black bellies and are joined on our shores by a host of other small waders in winter plumage that blends in with their muddy surroundings. Some are fairly scarce visitors, but a few come here in good numbers. If your Dunlin looks a bit stocky and has a pale stripe over the eye and a short straight bill it could be a Knot. They form large flocks on major estuaries having flown here all the way from Iceland and, incredibly, Arctic Canada. Also sporting a short straight bill is the much whiter Sanderling, an energetic bird of sandy beaches which chases the waves in and out looking for food. If your Dunlin is hanging around on seaweed-covered rocks, particularly along northern coasts, looks decidedly dark and has orangey-yellow legs it could be a Purple Sandpiper (*see Turnstone*). Finally, the Common Sandpiper is a similar-sized brown-backed, white-bellied, straight-billed bird which is generally found by freshwater and repeatedly bobs up and down as it walks (*see Dipper*).

# Snipe

## Key identification features

■ The Snipe is a brown bird of boggy places with a preposterously long beak.

■ Its plumage is subtly marked with light and dark feathering that provides excellent camouflage and its grey-green legs look short beneath its stocky white-bellied body. Pale feather fringes create buff lines down its dark brown back and its face is faintly striped, with lighter bands above and below the eye and along the top of the dark crown.

■ In flight the long bill is visible, as well as the stripes running down its back and the pointed wings have a thin tracing of white along the trailing edge.

## Where could you find one?

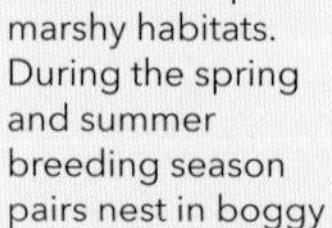

Feeding on worms and other invertebrates, the Snipe needs soft wet mud in which it can probe its bill, so is generally found in damp and marshy habitats. During the spring and summer breeding season pairs nest in boggy upland areas and damp tussocky pasture, particularly in northern Britain, as well as lowland marshes and meadows. In winter they are more widespread and turn up on flooded grassland, estuaries and wetlands.

## Is it easy to identify?

The Snipe is basically an extraordinarily long bill with a bird attached. The straight beak is proportionately lengthier than that of any other British species, sticking out in front like a conductor's baton. It is also a beautifully marked bird, in an understated way. The darker brown back and lighter front are embroidered with exquisite patterning of varied tones that break up its outline and blend in among the tangled grasses and muddy stretches it calls home. The challenge is getting a decent view and an encounter on a walk almost always takes you by surprise as an unseen bird explodes from the ground a few paces ahead and rapidly zigzags skyward delivering a couple of rasping croaks. You don't get time to take in much detail – perhaps the short tail, angled wings and that long bill, visible even when it climbs high and heads away.

## How likely am I to see one?

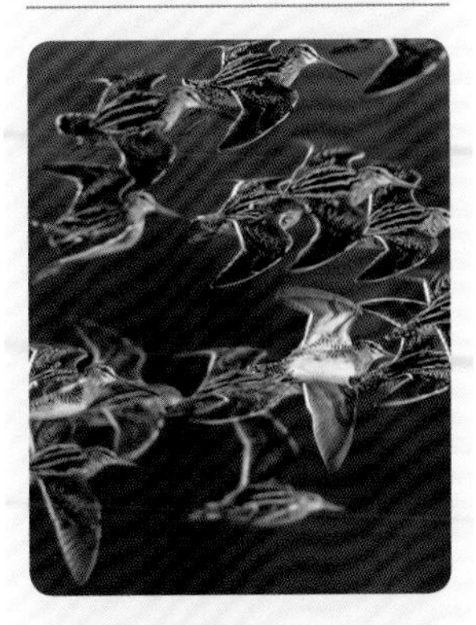

A flock takes flight

Around 75,000 pairs breed in Britain during the spring and summer, with the greatest concentrations in Scotland and the uplands of northern England. This is the time to witness the males' 'drumming' flight displays, especially at dawn or dusk. Over the colder months our resident population is joined by migrants from northern Europe, boosting numbers to over a million.

## What makes it special?

The Snipe is not the only wader to have an incredibly elongated bill. Other species, such as the Black-tailed Godwit and Curlew, have lengthy mud-probing beaks – but also the long legs to match. Snipe stand low to the ground making their outsized worming tool appear big for their body. It does the job though, being plunged down into the soft ground with a sewing machine action to snaffle up invertebrates that lie beneath the surface. There is more to a Snipe than just its beak, however. It is also known for the unique sound it makes when displaying over breeding territory, which is given the term 'drumming'. Much like the drumming of woodpeckers on trees, it is a mechanical rather than vocal noise, though generated in a very different fashion. A male Snipe rises high in the air then repeatedly dives down, spreading his tail so the stiff outermost feathers stick out almost horizontally, vibrating rapidly in the airstream to create a strange bleating sound. It has given rise to an assortment of names for the bird, including 'flying goat' and 'heather-bleater'.

The flight display known as 'drumming'



## Added interest

- The bill tip of the Snipe is touch sensitive and contains specialised nerve endings that enable it to detect the presence of prey as it probes in the mud.
- Snipe don't go in for confusing seasonal plumage changes like many other wader species of damp places, and look the same whatever the sex or time of year.
- At one time nesting pairs were far more widespread, but breeding has declined steeply over recent decades in lowland England and Wales as damp grassland fields have been drained.

## Not quite a Snipe?

The Snipe has a smaller and less common relative: the Jack Snipe. These shy birds look very similar, but have a shorter bill, a dark crown which lacks a central light stripe and clearer, creamy lines down the back. Around 100,000 visit in the colder months and they are most likely to be spotted when flushed on a walk, flying in a flatter trajectory than zigzagging Common Snipe. Another similar bird is the Woodcock – a larger, broader-winged species with marbled chestnut-brown plumage and dark stripes across the head, rather than lengthways like a Snipe. This is a widespread, secretive and largely nocturnal woodland resident, most easily spotted when disturbed or in the spring as males perform display flights over their forest territories.

Woodcock

# Lapwing

Female, with shorter crest

## Key identification features

■ The Lapwing is an attractive pigeon-sized wader with broad rounded wings and a wispy crest.

■ It is white below, dark on top and in good light the back has a greenish-purple metallic sheen. A black band runs across the breast and the white face has black markings near the base of the short bill, below the eye and on the steep forehead that sweeps back into a thin crest.

■ The legs are pinkish and there is a patch of orange feathering under the tail.

## Where could you find one?

Present in Britain all year round, the Lapwing breeds on open farmland, rough pasture, moors, meadows and marshes, particularly in northern England and Scotland. In the winter, flocks are mobile and widespread, gathering on lowland pasture, ploughed fields and coastal mudflats.

## Is it easy to identify?

No other bird looks quite like the Lapwing. It has a long crest that curves back from the top of the head in thin black strands which are visible even at a distance. Given birds in Britain generally don't go in for fancy head adornment, this flyaway fascinator is, in itself, a distinguishing feature. The Lapwing is also very conspicuous in flight: bold contrasting plumage alternates dark above and flashes of white beneath. It has a gentle expression, with large eyes used to spot insect prey and the feathers on the back have an attractive burnished purple-green gloss in sunlight. It adopts a characteristic stop-start style of hunting ground-invertebrates by sight – taking a few quick strides, pausing to look, then tilting forward to peck with the short bill. The kazoo-like call is distinctive and in the spring they perform noisy, tumbling flight displays over their territories.

## What makes it special?

Eccentric, endearing and in decline, the Lapwing is a bird to treasure and is among Britain's best-loved species. Always a joy to see and hear, this charismatic farmland and wetland favourite is sadly half as common as it once was. Numbers of breeding pairs have plummeted in recent decades and its future is a cause for concern. Lapwings are birds of open country that lay their eggs in a sparsely-lined scrape in the soil or short turf where they have a good view of any potential predator approaching. Once the clutch has hatched, parents lead their chicks to damp and insect-rich grassy areas nearby in which to feed. Changing agricultural practices mean that mixed farms providing a mosaic of habitats have largely given way to specialisation and intensification; rough wet grassland has been drained; pesticide-use has reduced insect prey and a switch to the sowing of cereals in the autumn rather than spring, in order to get a head start, results in crops growing too tall for nesting.

### How likely am I to see one?

Around 130,000 pairs of Lapwings breed in Britain, with the greatest numbers concentrated in northern areas. During the autumn and winter they gather in flocks and numbers in Britain are swelled by migrants escaping colder conditions in northern Europe, reaching 620,000 birds. Winter is the easiest time to see Lapwings, on farmland and wetland areas.

## Added interest

- Breeding pairs can be told apart as the male has a longer crest and the female has white mottling on her black bib.
- The collective noun for a group of Lapwings is a 'deceit', based on parents' devious tactics when protecting eggs and young. Predators are lured away by the birds feigning injury or pretending to defend a false nest site at a safe distance from the real one.
- Lapwing eggs were considered a culinary delicacy by the Victorians, though plundered in such numbers that the bird was eventually protected with its own Act of Parliament in 1926.

### Not quite a Lapwing?

The distinctive shape, pied plumage and flappy flight style of Lapwings set them apart. Other similar-sized black-and-white birds include the Oystercatcher, a coastal wader with a longish red bill and white up the lower back in flight (*see Oystercatcher*) and the Magpie, which has a long tail and white shoulders (*see Magpie*), but they are unlikely to be confused. Close relatives in the plover family, such as the Golden Plover, share the Lapwing's shape and feeding style, though lack the very dark back and crest (*see Golden Plover*). If it looks like a Lapwing, but is paler and browner in tone and only has a short crest, then it could be a young bird.

Juvenile Lapwing

# Golden Plover

Winter plumage

Breeding plumage

##  Key identification features

■ The Golden Plover is an attractive dove-sized wader with longish legs, a short bill, plump body and a small rounded head.

■ The back is speckled with yellow, dark brown and grey and in the spring and summer breeding season the front of the face, neck, chest and belly is black, fringed with a white border. In the autumn and winter the Golden Plover is much plainer, with a streaky warm beige front and whitish belly.

■ The underside of the pointed wings is white and in flight a thin pale line is visible running down the centre of the dark upper-wing.

##  Where could you find one?

There's gold in them thar hills... The Golden Plover breeds in the uplands, mainly in Scotland and northern England, nesting in open areas of moorland, blanket bog and grassland. Outside the breeding season it moves downhill to more lowland areas, including arable fields, damp pasture and saltmarshes and is widespread across most of Britain.

## Is it easy to identify?

In spring and summer the Golden Plover makes for a wonderful sight in smart breeding plumage. Dressed for the occasion, it looks like a black bird that has donned a shawl of glittering sequins. The dark feathering covers the front of the face all the way down to the belly, while an ermine-fringed 'cloak' spangled with gold, white and dark brown wraps around the rear of the head and over the back. One is more likely to spot Golden Plovers in plainer non-breeding plumage during the autumn and winter. At this time of year the bird is an understated buff brown, with a pale belly and subtle golden freckling on the back. They have a completely white under-wing, visible in flight or when stretching. When feeding, they cover ground in a stop-start manner, running a few steps then pausing to bend and pick up invertebrates.

## What makes it special?

While spring is the time to see Golden Plovers looking their most stunning, they are equally impressive when they ditch the formal wear in the autumn and join together in vast flocks. Swelled by winter migrants from northern Europe, these gatherings can be hundreds and even thousands strong. On the ground a Golden Plover hardly looks like a supreme flier, but once airborne it is streamlined and surprisingly fast, cutting through the air with falcon-sharp wings. Flocks race through the sky in tight formation, their white underwings flashing as they turn and sweep back and forth over roosting areas, taking an age before finally landing. They are wary birds and their large eyes give them a rather worried expression. They also have a melancholic call, a fluty whistle that is evocative of windswept grasslands and desolate moors. In the spring the Golden Plover can seem a plaintive-sounding and lonesome species of the barren uplands, but by winter, in simpler plumage, it becomes a gregarious bird capable of awe-inspiring aerial spectacles.

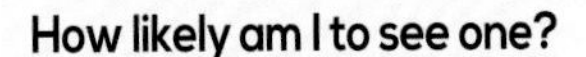

### How likely am I to see one?

During the spring and summer more than 35,000 pairs inhabit upland moors and grasslands, largely in the northern half of Britain. Over the colder months, numbers are boosted more than ten-fold by over-wintering migrants from Iceland and northern Europe. At this time of year, they can be seen across Britain, both inland and on coastal mudflats.

## Added interest

- The amount of black on the face and front of a Golden Plover depends on how far north it breeds, with those from Iceland being much darker than southernmost British birds.
- In the autumn and winter, flocks will mix with lapwings to feed in fields, though these congregations tend to attract gulls that persistently steal their worms.
- A debate among members of a shooting party in Ireland over whether a flying Golden Plover was the fastest game bird inspired brewery manager Sir Hugh Beaver to publish a book of superlative feats and facts in 1955 – The Guinness Book of Records.

### Not quite a Golden Plover?

If you are by the sea and you spot a Golden Plover whose spangled back looks more silver than gold, then it could be a Grey Plover. These close relatives don't nest in Britain, but more than 40,000 fly from their Arctic breeding grounds to spend the winter all around our coast. They generally spread out when feeding on sandy and muddy shores, such as estuaries, though may bunch up in flocks at high tide. Looking much like grey versions of Golden Plover, if a little sturdier and thicker billed, they can be told apart in flight as they have a white rump and jet black armpits. A great bird to identify.

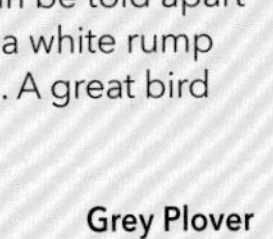

**Grey Plover**

# Seabirds

# Herring Gull

Juvenile

## Key identification features

■ The Herring Gull is a large, confident seabird, with white plumage and a neat silver-grey back. It has a strong, slightly hooked yellow bill marked with a spot of red near the tip and its legs and webbed feet are pink.

■ The wings have black ends with white spots and it has pale yellow eyes and a slightly fearsome expression. Males and females are alike.

■ Juvenile birds are the same size as adults, but are mottled brown and have a dark bill and dark brown band at the end of the tail. They acquire adult plumage through successive moults over four years.

## Where could you find one?

The Herring Gull is a common bird around our coasts and inland. They nest in noisy congregations on sea cliffs, islands and increasingly on rooftops. Intelligent and opportunistic, Herring Gulls hang around seaside towns looking for easy meals, follow fishing boats and are found at harbours, beaches, rubbish dumps, on fields, farmland, estuaries and reservoirs.

## Is it easy to identify?

The adult Herring Gull is a smart, familiar species, with clean white plumage, a uniform pale grey back and an assertive air. It flies effortlessly on broad wings and its evocative yelping cries provide the soundtrack to seaside holidays. However, it can be confused with other gulls (*see right*), so a few key features should be ticked off to be certain you are looking at a Herring Gull. It has pink legs and feet, not red or yellow and its yellow bill has a red spot near the tip, which is absent in some gull species. Its back is pale grey, not dark slate-grey or black and its black wing tips have small white patches. In winter adults have dusky streaks on the head and neck. Blotchy brown juveniles are not easy to separate from those of other similar-sized gulls.

## What makes it special?

The Herring Gull is the quintessential sight and sound of the British coast. However, this raucous and resourceful member of the gull family has earned itself something of a bad reputation in towns and cities where thousands of pairs now breed. Exchanging sea cliff nest sites for rooftops and fresh fish for the contents of bin bags, they have made themselves unpopular with their messy, noisy and occasionally aggressive antics, which range from dive-bombing people straying near their chicks to stealing chips from unwary holidaymakers. Yet despite thriving in built-up areas, Britain's wider population of Herring Gulls has collapsed over recent decades, to the extent that they are now included on the UK's Red List. The fact they are losing ground demonstrates that even relatively common species can never be taken for granted. Few birds create a sense of place like the Herring Gull – a seaside wouldn't be the same without them.

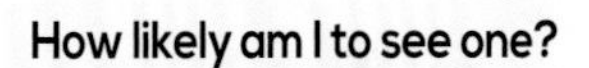

### How likely am I to see one?

This gregarious heavyweight seabird is hard to miss around our shores, given its size and loud calls. More than 130,000 pairs breed in Britain and 700,000 individuals spend the winter here, boosted by visitors from Scandinavia.

## Added interest

- The Herring Gull has mastered the art of breaking open mussels by dropping them onto rocks. However, flying high enough to ensure a shell cracks on landing runs the risk a rival gull at ground level will steal it.
- The red spot near the tip of the gull's bill acts a bit like a vending machine button – a tap on it by chicks prompts adults to regurgitate food.
- If you come across a Herring Gull on a flat grassy area apparently running on the spot with rapid Riverdance-style foot movements, it is engaging in a spot of worm charming. Drumming its feet on the turf is thought to imitate heavy rain and draw worms to the surface.

### Not quite a Herring Gull?

If your Herring Gull looks slightly smaller and has dark eyes and a yellow bill which lacks a red spot, then it could be a Common Gull or Kittiwake. The Common Gull shares the white patches on the black wing tips, but has greenish-yellow legs. The Kittiwake has black legs and solid black wing tips (*see Kittiwake*). If your Herring Gull has a dark back, then it could be either a Lesser Black-backed Gull or a Great Black-backed Gull. The Lesser Black-backed Gull, numbering over 100,000 pairs in Britain, has a dark grey back and yellow legs. The Great Black-backed Gull is larger and blacker across the back and has pale pink legs and around 17,000 pairs breed in Britain, mainly in western areas.

**Lesser Black-backed Gull**

# Black-headed Gull

Winter plumage

Breeding plumage

##  Key identification features

■ The Black-headed Gull is our smallest common gull, with thin pointed wings and fine features.

■ Both males and females are white with a silvery grey back and their legs and beak are red. From late winter until early summer the front of the head is dark brown, while for the rest of the year this 'mask' disappears, save for a smudge behind the eyes.

■ In flight, the outer half of the wing has a noticeable white front edge and a tracing of black behind. Young birds have some messy brown feathering across the wings and back, as well as a black band across the tail end.

## Where could you find one?

This gregarious bird can be found year-round across most of Britain. Noisy nesting colonies are situated close to the sea and inland, on marshes, shingle spits, dunes and beside upland pools and lowland water bodies, including islands in reservoirs and flooded gravel pits. Outside the breeding season the gulls are more widespread, feeding in flocks on a range of habitats including wetlands, beaches, landfill sites, park lakes and farmland.

##  Is it easy to identify?

In its finest breeding plumage, the Black-headed Gull is a handsome and distinctive bird. From a distance the front of the head looks black, as its name suggests, but good views reveal this to be a deep chocolate brown. The dark eyes are accentuated by a thin white crescent of 'eyeliner' and it has red legs as well as a red beak. Things get trickier outside the spring and summer breeding season however, when it gradually loses its dark head colouring – save for a couple of sooty fingerprint-like markings behind the eyes. The Black-headed Gull has a brushstroke of black skirting the rear of the outermost flight feathers and, most eye-catching of all, a glinting blade of white along the front edge. Consider it a feather in your cap to recognise a Black-headed Gull in its white-headed winter guise!

## How likely am I to see one?

Park lakes attract Black-headed Gulls

Around 130,000 pairs nest in Britain during spring and early summer and may be spotted in colonies at freshwater sites and beside the coast. In the colder months, winter migrants from Scandinavia, northern Europe and Iceland come to Britain and push up numbers to over two million birds, with flocks a common sight in rural, urban and seaside areas.

## What makes it special?

Britain's Black-headed Gulls could be dubbed 'landgulls' as much as 'seagulls', given they are perfectly at home in town parks, on sports pitches, arable fields and rubbish dumps, as well as around the coast. Although large numbers of northerly breeders cross the sea to escape freezing conditions in winter, boosting our mostly stay-at-home resident population, they don't make a living on the storm-tossed oceans like the Kittiwake or Fulmar. Our very terrestrial Black-headed Gulls are more likely to follow a farmland plough than a deep-sea trawler. They are opportunists and have thrived over the last century, colonising newly-created habitats such as reservoirs and flooded gravel pits and taking advantage of novel food sources, from landfill waste to bread handouts at urban lakes. Confident and resourceful, filling the air with their screeching cries, they are quick to snatch food from ducks and other birds and agile enough to catch flying ants and insects on the wing.

## Added interest

- Seabirds can be extremely long-lived and the Black-headed Gull is no exception. One young bird ringed in Worcester in December 1980 was found dead in the Netherlands more than 32 years later.
- A Black-headed Gull featured in Richard Adams' novel Watership Down, the character's name Kehaar closely matching the species' strident and grating call.
- Legal protection in the 1860s helped curb the unsustainable slaughter of the gulls and raiding of nests for eggs, harvested in large numbers and sold as a delicacy to restaurants.

## Not quite a Black-headed Gull?

In the winter when Black-headed Gulls lose their dark coffee-coloured head plumage, they bear a resemblance to the similar-sized Kittiwake and Common Gull. If it is spring or summer and your Black-headed Gull is stocky looking and has a full dark hood, pale largely unmarked wings and a thicker red bill, then you might have struck lucky and spotted a Mediterranean Gull. This is a relative newcomer to Britain, breeding mainly in southern and eastern England, often alongside Black-headed Gulls and is more widespread in winter when it numbers 1,800. Finally, if your Black-headed Gull has a black cap that sits on top of the head, rather than covering the front of the face, then it could be a kind of tern (*see Common Tern*).

Mediterranean Gull

# Kittiwake

In flight

## Key identification features

- The Kittiwake is a neat and attractive seagull which is smaller than a Herring Gull, but shares the white plumage and grey back.
- It has an unmarked yellow bill, dark eyes, quite short black legs and the wings are tipped with triangles of solid black.
- The sexes look alike and in the winter develop dark grey smudges behind the eyes. Juvenile birds have wide black zigzags across the wings, a dark collar and black tip to the tail. Kittiwakes get their name from their raucous *kitti-wake* call.

## Where could you find one?

Kittiwakes are common breeding birds around our rocky shores, particularly in Scotland and north-east England. Their traditional colonies range from hundreds of pairs to tens of thousands and nests are built on sea cliffs as well as on the ledges of bridges and buildings near the coast. In the autumn and winter they head to sea and can be spotted flying offshore, but virtually never stray inland.

## Is it easy to identify?

Juvenile

Gulls don't always make identification easy given we have a few similar-looking species. However, the Kittiwake is at the smaller end of the scale with a light grey back. The Kittiwake differs from the Herring Gull and the Common Gull in two main ways: the short legs are black and, in flight, the wing tips are pure black, as if dipped in ink. These key factors remain the same even in winter. The juvenile Kittiwake has a black line that runs along the front outer flight feathers then angles back towards the body, as if an 'M' has been drawn with a thick marker pen across the open wings. This bold pattern gradually fades, along with its black collar and tail tip. Coastal breeding colonies are noisy and pairs repeatedly cry out their name *kitti-wake, kitti-wake*, though the gulls are largely silent the rest of the year.

## What makes it special?

A large Kittiwake colony in spring is a spectacular sight as thousands of birds crowd cliff sides, filling the air with their cries. They nest close together on precarious ledges, building a cup-shaped mound of mud, seaweed and grass in which they typically lay two eggs, taking turns with incubation. Safely out of reach of most predators, the young grey-and-white chicks don't need camouflaged brown plumage, but are at risk of falling so have strong sharp claws and remain in the nest until they can fly. Come the early autumn, Kittiwakes head off to sea and may not set foot on land until they return six months later. They are our most ocean-going gull: graceful maritime wanderers that would never demean themselves by scavenging at rubbish dumps or stealing holidaymakers' chips. Some young adventurers can drift for at least three years before trying to start a family. While a cacophonous Kittiwake colony is one of our most impressive avian spectacles, this hardy species' endurance at sea is also remarkable. Next time freezing winter storms are pounding at your window, think of our Kittiwakes riding out the gales mid-Atlantic.



### How likely am I to see one?

The Kittiwake is not only the world's most abundant gull (with a population of over 17 million) it is also our most common breeding gull, though has suffered recent declines. Around 370,000 pairs nest in Britain and the best time to see them is in the spring and summer before colonies disperse.

## Added interest

- The plundering of eggs and shooting of Kittiwakes for sport or use in the Victorian fashion industry, led to the first wildlife legislation of its kind, the Sea Birds Preservation Act of 1869.
- In the breeding season the majority of successful pairs reunite at the nest site, calling loudly to each other and reaffirming their bond by touching bills before sharing the duties of repairing the nest.
- Kittiwakes travel widely, feeding on fish at the surface of the sea and have been spotted by Arctic expeditions as far north as the North Pole.

### Not quite a Kittiwake?

The two grey-backed gulls that look most similar to the Kittiwake are the larger Herring Gull, which has pink legs, white spots on the black wing tips and a red spot on the beak (*see Herring Gull*) and the similar-sized Common Gull. Common Gulls share features with both, having the white wingtip spots of the hefty Herring Gull and the dark eyes and yellow beak of the Kittiwake, but they also have yellowish legs. They breed in northern Britain and are abundant in the winter, both around the coast and more often inland, with visiting birds from Iceland and Scandinavia boosting numbers to 700,000.

Common Gull

# Fulmar

## Key identification features

- The Fulmar is a gull-like seabird with a stiff-winged flight action and prominent tube-shaped nostrils.
- The body plumage is white and the back and top of the wings are mottled blue-grey, while the upperside of the tail is pale grey.
- The Fulmar generally glides when flying, holding its wings out straight and rigid. The head and thick neck are white, the eye is dark and the bill has a pale yellowish hooked end and grey nostril tubes on top.

## Where could you find one?

A species of the open seas, the Fulmar can be spotted all around our shores, though never inland like some gulls. Pairs nest on rocky cliffs, particularly in northern and western Britain, occasionally on suitable coastal buildings and are a standard feature at large seabird colonies.

## Is it easy to identify?

At first glance the Fulmar looks like a gull, with white plumage and a grey back, but as soon as it takes to the wing its closer links with albatrosses become evident as it cuts through the air on narrow blades stretched out straight and rigid. The ocean-going Fulmar effortlessly rides the winds along cliff faces and the updraughts created by waves, virtually skimming the water as it scores a path across the sea. In the right conditions, this super-efficient flier's masterful use of air currents means it can keep airborne and travel vast distances with scarcely a wingbeat. It has quite a robust body, large dark eyes and a lighter patch in the outer flight feathers, though no black wing tips. The Fulmar also has an unusual-looking beak that has long nasal passages and shares this feature with other 'tubenoses', including shearwaters, petrels and albatrosses.

## What makes it special?

When storms whip the seas into a rage, nothing looks more in its element amid the foam-crested waves than a Fulmar. Skimming the peaks and troughs with scarcely a wingbeat, this tough seabird is a wonder to watch, gliding low over the muscular swell and switching from white belly to grey back as it wheels in the wind. As soon as young birds leave the nest, they wander the North Atlantic for several years, catching fish at the surface, feeding on the floating carcasses of marine creatures and scavenging discarded offal from trawlers. Only when they are aged about four or five will they return to breeding sites and begin trying to secure a mate for life and a safe nesting ledge. Fulmars are long-lived birds and can survive into their 30s and 40s. And while they begin life as ocean voyagers, pairs become surprisingly attached to their nest sites. They range widely in search of food, spending time at sea in the late summer after the breeding season, but are soon back defending their precious square of turf or rocky recess, and at traditional breeding areas a few may be present even during the winter.

### How likely am I to see one?

Fulmars are common and widespread, with more than 500,000 pairs breeding in Britain. They are most easily viewed at sea cliff colonies, especially during the spring and summer, and also on boat trips which may attract feeding flocks.

## Added interest

- At one time Fulmars only bred in Britain on the remote Outer Hebridean islands of St Kilda, where they were harvested for food, oil and feathers, but colonised Shetland in 1878 and then spread south all around our rocky coasts during the twentieth century.
- 'Tubenoses', such as Fulmars, have a good sense of smell, enabling them to track down food at sea over impressive distances, and also excrete excess salt through the nostrils.
- Fulmars have a nasty surprise in store for those who approach their nest too closely: they eject the oily contents of their stomach over intruders – a habit from which their original Norse name, meaning 'foul gull', derives.

### Not quite a Fulmar?

If your Fulmar is flying low over the sea in the typical stiff-winged fashion, but is slim-looking and black, instead of grey, above – including on the top of the head and back of the neck – then it could be a close relative, the Manx Shearwater. More than 300,000 pairs nest in burrows on offshore islands along western coasts and they can be spotted on boat trips and from headlands. Always a sighting to savour, they are supreme ocean wanderers, migrating thousands of miles to the coast of South America in winter.

Manx Shearwater

# Common Tern

In flight

##  Key identification features

■ The Common Tern is a black-capped, white seabird with a slender build, pointed wings and a forked tail.

■ Smaller than our resident gulls, it has a white body, light grey back and the top half of its head is black. The legs are orangey-red, as is the sharp beak, which has a black tip. It plunge-dives for fish and has a harsh shrieking call.

##  Where could you find one?

The Common Tern only comes to Britain for the spring and summer, breeding in colonies around the coast on shingly and sandy shores. Increasing numbers also nest inland, especially in southern and eastern England, on islands in reservoirs, rivers or flooded gravel pits, including on specially-made rafts where they are safe from predators. They can be seen at noisy colonies, fishing inshore or over open freshwater.

##  Is it easy to identify?

At first glance terns look like slim gulls, given their silvery-grey backs and white undersides and yet they are much more buoyant in the air, with a swift, springy flight action. Of the tern species that nest in Britain, the Common Tern is the most widespread and has smart pale plumage, a neat jet black cap and matching orangey-red legs and beak. Its twin, the Arctic Tern, is virtually identical, though mostly breeds in large colonies in Scotland. Those unable to tell them apart in the field simply merge their names and record them as 'comic' terns. It's no laughing matter trying to tell one from the other, but there are subtle differences: the easiest one being that the slightly longer-legged, shorter-tailed, darker-winged Common Tern has a black tip to its red bill. It is also the most abundant tern in England, with increasing numbers nesting inland at suitable freshwater sites.

## What makes it special?

Common Terns are beautifully sleek and streamlined birds, from their pointed beaks and dapper black caps to their forked tails. They cut through the air on scalpel-sharp blades and are expert at catching fish, hovering several metres up then plunge-diving to snatch small species such as sprats and sand eels from just beneath the surface. Offerings of freshly-caught fish help males to secure mates, demonstrating their prowess as providers while ensuring females are well fed ahead of the energy-sapping process of egg-laying. And it is on their breeding grounds that terns are at their boldest and brashest best. Graceful and chic they may appear, but nesting birds are capable of making an almighty racket, piercing the air with strident calls as pairs keep in touch, neighbours bicker and adults join forces to mob approaching intruders. A colony in full voice is a wonderful raucous celebration of life.



### How likely am I to see one?

Around 10,000 pairs of Common Terns breed in Britain, they arrive in the spring and depart by autumn for warmer waters off tropical Africa. During the breeding season they can be spotted at and around scattered traditional nesting sites on beaches, or at some inland lakes, where they lay their eggs in a scrape in the ground. If you see a bird carrying a fish it is either a male wooing a mate or parents ferrying food for their chicks.

## Added interest

- During the nineteenth century, large numbers of Common Terns were killed and their feathers and wings used to adorn Victorian hats, before laws were introduced to protect seabirds from persecution.
- Common Terns are among the longest lived of our birds, with one individual ringed in Northumberland in 1963 being spotted alive and well near Liverpool in 1996, some 33 years later.
- Rather than brave cold and stormy northern winters, Common Terns head for western and equatorial Africa, while Arctic Terns undertake the longest migration of any animal on the planet, travelling to Antarctica and back.

### Not quite a Common Tern?

If your Common Tern has a blood red beak with no black tip, pale ends to the upper-wings and short squat legs, then it could be an Arctic Tern. Numbering more than 50,000 pairs, they mainly breed on the coast and islands of Scotland, as well as northern England and north Wales. If your Common Tern looks particularly gull-sized and short tailed and has a dark bill and legs, as well as spiky feathering at the back of the black cap, then it could be a Sandwich Tern. These pale and powerful plunge-divers are scattered in coastal breeding colonies around Britain. Finally, at the other end of the size scale is the Little Tern, a coastal species with a yellow bill, short forked tail and whirring wingbeats - a scarce sighting, with fewer than 2,000 breeding pairs nationwide.

Arctic Tern

# Gannet

In flight

## Key identification features

■ The Gannet is a large and impressive seabird with a long body and wings, pointed tail and formidable beak.

■ The plumage for both males and females is pure white, flushed orangey-yellow on the head and neck and the ends of the wings are black. Pale eyes are set against a black surround and ringed with pastel blue, the bill is light grey.

■ Juveniles are speckled brown and become whiter with age, taking four years to attain full adult plumage. Gannets fly strongly, gliding frequently and plunge-dive from height when feeding.

## Where could you find one?

These ocean-roamers only come to land in the spring and summer to breed, building guano-plastered nests of seaweed and marine debris on rocky islands and cliffs where colonies are safely out of reach of ground predators. There are a number of well-established 'Gannetries' in Britain, ranging from Bass Rock in Scotland's Firth of Forth and RSPB Grassholm in Wales (both of which can be visited by boat) to the mainland RSPB Bempton Cliffs reserve in Yorkshire. Feeding and non-breeding Gannets are widespread and although birds head to sea in the autumn and winter they can be spotted year-round offshore.

## Is it easy to identify?

The Gannet is the biggest of our seabirds and dazzling white. Goose-sized, strong-necked and dagger-beaked, it lacks any gull-grey across its upper-side and the ends of the long wings are solid black. The body is missile-shaped and streamlined, head tapering into the solid spear of a bill and tail feathers narrowing to a point. Gannets look strong in flight, powering their way purposefully over the sea on a six-foot wingspan and close views reveal an attractive wash of apricot plumage on the head and neck, pale staring eyes and dark outlining on the blue-grey bill. They are best known for their plunge-diving, plummeting headlong from a height to catch fish and a feeding flock is instantly recognisable. In fact, it is believed their black-and-white plumage is intended to be conspicuous, attracting others to join the fishing frenzy, disorientating and exhausting shoals and making panicking prey easier to catch.

## What makes it special?

Spectacular when feeding, remarkable when breeding, the Gannet is a striking species and we are lucky enough to have the majority of the world's population living around the British Isles. The sights, sounds and smells of a large colony during the nesting season are not to be missed. From a distance the sea cliffs look to have been dusted with hailstones: every ledge and flat section of turf dotted with a multitude of evenly-spaced white adults and the surrounding rocks spattered with guano. The sky is busy with birds and the cackling clamour of bickering neighbours and courting couples fills the air – as hectic as an airport terminal! When feeding, these high-divers scan the water from about 10-30 metres up. With a target in their sights they drop from the sky beak-first, outstretched wings closing up behind them like a folded umbrella as they pierce the water at speeds exceeding 60mph. Translucent membranes protect the eyes from the impact and the shut beak prevents water getting into nostrils which open inside the mouth. The headlong plunge enables these buoyant birds to either snatch fish after entering the water, gulping them down immediately, or torpedo deep enough to swim after prey using their webbed feet and wings. A blizzard of Gannets bombarding a shoal of fish is among our most breathtaking wildlife spectacles.

### How likely am I to see one?

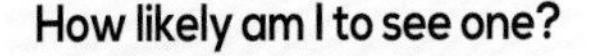

Around 220,000 pairs of Gannets nest at fewer than 20 breeding sites around Britain, with the largest colonies having tens of thousands of birds during the spring and summer. Even if you don't pay a visit to one of these scattered, often remote, locations, you can still see Gannets anywhere around our coast.

## Added interest

- Pairs reunite in early spring at their established nest site and regular meet-and-greet rituals involve standing chest to chest, pointing their bills skywards and wagging their heads from side to side.
- Gannet females lay a single egg and both parents incubate it beneath their webbed feet, using an increased flow of warm blood through their toes to maintain its temperature.
- Fully-grown chicks are abandoned by their parents, but are able to live off ample fat reserves until they eventually leave the nest and master the art of feeding themselves.

## Not quite a Gannet?

If your Gannet is brown, or partially so, then it is a young bird. Juveniles start out a dull chocolate colour freckled with white and look like they might be another species altogether, but for their size and shape. Over successive moults more white appears in the plumage, firstly on the head and belly then finally the back, until by four years of age they acquire the smart adult appearance and are ready to breed. Young Gannets might be confused with young gulls, which are mottled brown, or possibly the Great Skua, a dark seabird with white wing patches.

**Juvenile Gannet**

# Cormorant

## Key identification features

- The Cormorant is a large black bird with a thick sinuous neck, webbed feet and a yellow and white bare area at the base of its long beak.
- In the early spring, adults have white thigh patches and some fine silvery streaking on the head, though this gradually disappears during the summer. They dive for fish and can often be seen perched in a prominent spot drying their outstretched open wings.

## Where could you find one?

This resident fish-eater can be found around our coast throughout the year. Cormorants breed in colonies of varying sizes on rocky shores, sea cliffs and islands and are more widely distributed during the winter. In recent decades pairs have spread inland, nesting near freshwater lakes, rivers and reservoirs in England.

## Is it easy to identify?

This large, long-necked and raggedy black bird of coasts and lakes is hard to miss. Its plumage has an attractive glossy sheen, with scaly bronze feathering on the back, a rather flat-topped head and a long, straight, hook-tipped beak. The base of the bill has a bare area of orangey-yellow skin surrounded by white – resembling a fried egg – and its turquoise eyes add to a slightly sinister expression.

Cormorants fish relatively close to shore and perch in full view to dry their broad wings. They may be confused with the very similar Shag which has a steep forehead, so its bill looks to have been stuck onto the front of its face, while the Cormorant's forehead tapers into the beak as one solid unit (*See Not quite a Cormorant?*). In flight the Cormorant can appear a bit like a goose, but has a longer tail and also glides.

## What makes it special?

There is something wonderfully primitive about a Cormorant, from the weathered face and webbed feet to the serpentine neck and loose-fitting feathers. Holding their wings out to dry after diving for fish, they can look like visitors from the Jurassic era: part-reptile, part-bird and seemingly ill-equipped for flight. And yet they are strong in the air, as well as expert swimmers. Instead of waterproof plumage, Cormorants have fairly water-permeable feathering, which reduces buoyancy and enables them to efficiently dive metres below the surface, propelling themselves with their webbed feet and remaining submerged for around half a minute at a time. As a result they spend periods of the day perched on rocks, posts or trees airing their soggy outstretched wings between underwater forays. Cormorants can put on a show when feeding, as larger catches are tackled at the surface, which makes for fascinating viewing if you spot one battling to swallow a plaice or overpower a slippery eel. However, their impressive hunting skills can bring them into conflict with angling interests and they have been persecuted in the past.



### How likely am I to see one?

This is a fairly common and widespread species around the coast, though is increasingly found at fish-rich waters inland. More than 8,500 pairs nest in Britain and during the winter the population is boosted to over 35,000 birds as migrants from northern Europe join our mainly stay-at-home residents on inshore waters and lakes.

## Added interest

- The ancient Chinese practice of fishing with trained Cormorants - their throats restricted by a band to prevent them swallowing large prey - never caught on in Britain, though King James I kept them for this purpose in London and appointed a royal keeper.
- Cormorants began breeding at inland freshwater sites during the 1980s and the droppings produced by colonies have been known to kill the trees in which they roost and nest.
- Liverpool's famous Liver Bird is believed to be a Cormorant, however its more raptor-like depiction on Liverpool Football Club's crest may be closer to the city's original royal emblem, which is likely to have portrayed an eagle.

### Not quite a Cormorant?

If your Cormorant looks decidedly brown and pale-fronted then it could be a juvenile, which takes a couple of years to attain the dark plumage of an adult. If, on the other hand, your Cormorant is on the small side, sleek and slender and has a steep forehead and thin bill then it could be a Shag. Unlike the Cormorant, this coastal bird is rare inland, breeding in colonies on rocky coasts mainly in the north and west. With over 27,000 pairs, Britain is home to more than a third of the world's population. In spring the Shag's black plumage has a greenish gloss and they sport a tufty crest at the front of the head, which helps with identification.

Shag

# Guillemot

SEABIRDS

## Key identification features

- The Guillemot is a neat and streamlined seabird with simple plumage: dark brown on top and white beneath.
- The chest is broad, the webbed feet set far back on the body and the head tapers smoothly into the long, narrow bill, giving it a pointed profile. In winter, the brown feathering on the neck and sides of the face turns white, save for a streak behind the eye.
- Guillemots typically stand upright on land and fly fast and low over the sea on short and narrow wings. Some more northerly individuals sport white 'spectacles' encircling each eye and running back in a thin line, and are known as 'bridled' birds.

## Where could you find one?

Most of the year Guillemots are at sea, but return to land to breed on rocky coasts, mainly in north and west Britain. They are gregarious birds, nesting on the kinds of steep-sided cliffs and offshore islands that are home to a variety of seabirds and can be seen in spring and summer at large colonies and on the surrounding sea. In the colder months they move offshore, but may be spotted from headlands as they pass by.

## Is it easy to identify?

The Guillemot is a member of the 'auk' family of seabirds and is the largest and most common of the four species that breed in Britain - the others being the Puffin, Razorbill and Black Guillemot. Auks are sometimes nicknamed 'penguins of the north', given their pied plumage, clumsy waddling on land and superb swimming abilities, using their wings to propel themselves underwater. However, unlike penguins our auks have not lost their ability to fly. The Guillemot is perhaps the most penguin-like, with its upright stance, narrow pointed bill and smart white front. Its upperparts are velvety chocolate brown, but in poor light this plumage looks black - the same colour as the very similar Razorbill. Telling these two apart can be tricky. The main difference is the shape of the beak - a pointed dagger for the Guillemot and a blunt-ended breadknife for the Razorbill (*see Not quite a Guillemot? right*).

## How likely am I to see one?

A 'bridled' bird

Our rocky coasts are home to around 900,000 pairs of Guillemots. Traditional colony sites are a hive of activity in spring and summer and are the best places to encounter this species in large numbers. They may also be spotted on ferry crossings or boat trips which provide sea-level views of them diving from the surface after fish.

## What makes it special?

Guillemots live life on the edge, quite literally, as they raise chicks on the narrow ledges of precipitous sea cliffs in densely-packed gatherings. They choose some of the most precarious breeding sites, opting for the kinds of towering rock faces and thin ridges that few other birds would dare to use. A pair's single egg may be laid on a flat sill just a few inches wide and is pear-shaped so that it rolls in a circle if knocked, making it less likely to plummet straight off the side. Such exposed nest sites are certainly safe from land predators, but not marauding gulls and Guillemots seek safety in numbers, huddling shoulder-to-shoulder with their neighbours. Their large and noisy congregations make for an impressive spectacle, though life on the crowded crags is fraught with danger for newly-hatched chicks. The ocean is the safest place for these adept swimmers, so at about three weeks of age, before they have even grown full flight feathers, the young birds launch themselves from the cliffs into the water far below and head out to sea. They don't do this alone, however. Their devoted dads are with them all the way, escorting the juveniles on their daredevil plunge and accompanying them until they are old enough to fend for themselves.

## Added interest

- The large and pointed eggs laid by Guillemots have a water-repellent shell that helps keep them clean, as well as unique patterns which enable parent birds to recognise their own in congested colonies.
- After the breeding season, ocean-going Guillemots moult their flight feathers all in one go, so are rendered flightless for several weeks until the replacements grow.
- Guillemots can hold their breath for over a minute and swim down tens of metres to hunt fish, with the deepest dive recorded being 180 metres – nearly twice the height of Big Ben.

## Not quite a Guillemot?

If your Guillemot looks black on the head, neck and back, rather than dark brown, then it could be a Razorbill. These close relatives are widespread seabirds of similar coastal habitat, but less numerous, numbering around 110,000 pairs. Like the Guillemot, they are present inshore during the spring and summer breeding season and can look very similar, but have a deep blunt beak with a thin white line across the top and side. And if your Guillemot is small and lacks a white front, but instead has a white patch on the side of the wing, then it is a Black Guillemot, a species mostly found in Scotland (*see Puffin*).

Razorbill

# Puffin

In flight

## Key identification features

■ The Puffin is a distinctive and characterful seabird with a compact build and colourful beak.

■ Males and females are alike and have simple body plumage: black on the back and white on the front. The legs and webbed feet are bright orange-red and the round head is greyish-white on the sides, with black on the top and around the neck. Dark eyes are encircled by red skin and set within triangles of grey-blue, while the bill is banded in red, pale yellow and dark blue and has a yellow-orange fold at the base where the two halves join.

■ In flight, on narrow whirring wings, Puffins look blunt at the beak end and short-tailed.

## Where could you find one?

During the breeding season, Puffins gather on grassy sea cliffs and islands where they nest in sheltered crevices between rocks and in burrows, either excavating their own or taking over those dug by rabbits. Colonies are scattered around the coast at suitable sites from south-west England and Wales to the north-east and northern Scotland. They are present from April to August and spend the rest of the year at sea.

## Is it easy to identify?

Photogenic star of postcards, calendars and magazines, the Puffin is one of our most familiar and recognisable birds. But while these charming and idiosyncratic coastal-dwellers are a doddle to identify from pictures, they may be overlooked in the wild by those who have never seen one in real life before, given they are much smaller than expected: roughly the size of a pigeon, though with a stouter body. Whizzing over the waves at a distance, when the bill colours are impossible to discern, they look like a little snub-nosed black-and-white species of no great consequence, even though the pale face and black neck collar are distinguishing features. However, once you get a better view, or adjust your sense of scale when scanning busy bird-covered cliffs, you are rewarded with a wonderful moment of recognition. They steal the show and become favourites in an instant.

## How likely am I to see one?

Puffins nest in colonies

The Puffin is a relatively common seabird and more than 580,000 pairs nest around Britain. Given that numbers are largely concentrated at traditional breeding locations, the easiest way to enjoy sightings is by visiting an established colony, with regular boat excursions operating at many places.

## What makes it special?

Few birds are as endearing as the pint-sized Puffin. It is impossible not to like them. The combination of small tubby frame, large round head, simple smart plumage, waddling walk, comical beak and slightly forlorn expression gives these unique seabirds irresistible appeal. Much of their attraction derives from their amusing human-like qualities, especially when they stand upright outside their nest burrows on turf-covered sea cliffs gazing out to sea like proud homeowners enjoying the view. Their crisp white front and black back suggest formal wear – though given an eccentric twist with the addition of orange legs, a bizarre beak and clown-like eye 'make-up'. Puffins are much tougher than they look, spending most of the year far out at sea. They only come to shore at traditional colony sites during the spring and summer to breed. Pairs

raise a single chick in the relative safety of a metre-long burrow and provide impressive quantities of sand eels and other small fish grasped neatly in their wide-hinged and serrated beaks. They use their tongue to hold catches in place as they chase and seize further prey underwater, skilfully accumulating average hauls of around a dozen, with the greatest beak-load recorded being a staggering 62 fish.

## Added interest

- The striking beak results from the annual growth of colourful outer plates, which are shed at the end of the breeding season leaving the bill looking much plainer.
- Older Puffins have more grooves in their sizeable bills and can live to an impressive age – one Scottish individual ringed as a nestling in 1975 was found alive and well nearly 37 years later.
- While highly social at coastal nesting sites, it is believed our Atlantic Puffins go their separate ways during the colder months, roaming the storm-tossed seas alone and far from land.

## Not quite a Puffin?

Seen well, the Puffin is unmistakable, though at a distance it could be confused with two other members of the auk bird family which share the dark back and white front: the Guillemot and Razorbill (*see Guillemot*). Both these species also have white on the face in winter, as does the fourth of our breeding auks, the Black Guillemot.

This species is a similar size to the Puffin and a handsome bird in the summer: black all over with eye-catching white wing patches and red feet. Almost all of our 20,000 pairs of Black Guillemots are found in Scotland and they can be seen inshore around rocky coasts – a great species to spot.

Black Guillemot

# The UK Red List for birds

All the main species featured in this book are common and widespread, and yet a number have suffered worrying declines and are included on the UK's Red List of birds of conservation concern.

The Red List is compiled every six or so years by the RSPB and other leading bird organisations, including the British Trust for Ornithology (BTO). It includes species that have, in simple terms, more than halved in number or range in recent decades, suffered significant historical declines or that are internationally threatened.

When the Red List was first drawn up in the 1990s, a total of 36 birds qualified for inclusion. By 2002, a further four species were added, while the 2009 figure jumped to 52. The latest *Birds of Conservation Concern* (BoCC4) report was published in 2015 and currently totals 67 – more than a quarter of the 244 regularly occurring species that were assessed.

The roll call of decline spans everything from scarce breeders, such as the Golden Oriole and Hen Harrier, to common garden visitors, like the House Sparrow and Starling, whose populations have plummeted. Other Red List birds include the Puffin, Cuckoo, Herring Gull, Lapwing, Yellowhammer, Curlew and Skylark.

The *Birds of Conservation Concern* report also places species in amber and green categories – a 'traffic light' system where red is the highest conservation priority, with species needing urgent action; amber is the next most critical group, followed by green. The BoCC4 Red List and Amber List are shown opposite.

Factors causing declines in bird populations range from habitat changes and persecution to the effects of climate change. Despite an increase in the number of our Red List species, it is not all doom and gloom. Conservation efforts have been rewarded as priority species such as the Bittern and Nightjar have bounced back, shifting from red to amber status, while the green list of birds least at risk has increased.

It is thanks to the surveys of conservation charities and volunteers, coupled with a long tradition of recording wildlife sightings, that we can identify the fluctuating fortunes of our fauna. This, in turn, enables the RSPB and other conservation organisations to use limited resources most effectively to help threatened species.

As the RSPB's Conservation Director Martin Harper explains: 'When we have diagnosed the problem, identified solutions, and when conservation action is targeted and adequately funded, we can bring species back from the brink.'

We can all do our bit to help birds and other species of conservation concern in the UK. Joining organisations such as the RSPB supports efforts to help threatened species. You can also take part in volunteer surveys to gather valuable data, or find ways to make outdoor spaces more wildlife-friendly – from putting up nest-boxes in the back garden to planting pollen-rich flowers or creating a pond.

**Puffin, a Red List species, on Skomer Island, Pembrokeshire, Wales**

## BoCC4 Red list

| | | | |
|---|---|---|---|
| White-fronted Goose | Ringed Plover | Golden Oriole | Nightingale |
| Pochard | Dotterel | Red-backed Shrike | Pied Flycatcher |
| Scaup | Whimbrel | Willow Tit | Black Redstart |
| Long-tailed Duck | Curlew[a] | Marsh Tit | Whinchat |
| Common Scoter | Black-tailed Godwit | Skylark | House Sparrow |
| Velvet Scoter | Ruff | Wood Warbler | Tree Sparrow |
| Black Grouse | Red-necked Phalarope | Grasshopper Warbler | Yellow Wagtail |
| Capercaillie | Woodcock[a] | Savi's Warbler | Grey Wagtail |
| Grey Partridge | Arctic Skua | Aquatic Warbler | Tree Pipit |
| Balearic Shearwater | Puffin | Marsh Warbler | Hawfinch |
| Shag | Roseate Tern | Starling | Linnet |
| Red-necked Grebe | Kittiwake | Ring Ouzel | Twite |
| Slavonian Grebe | Herring Gull | Fieldfare | Lesser Redpoll |
| White-tailed Eagle | Turtle Dove | Song Thrush | Yellowhammer |
| Hen Harrier | Cuckoo | Redwing | Cirl Bunting |
| Corncrake | Lesser Spotted Woodpecker | Mistle Thrush | Corn Bunting |
| Lapwing | Merlin | Spotted Flycatcher | |

a - species on the Amber list previously, g - species on the Green list previously.

## BoCC4 Amber list

| | | | |
|---|---|---|---|
| Mute Swan[g] | Manx Shearwater | Common Sandpiper | Stock Dove |
| Bewick's Swan | European Storm Petrel | Green Sandpiper | Tawny Owl[g] |
| Whooper Swan | Leach's Petrel | Spotted Redshank | Short-eared Owl |
| Bean Goose | Gannet | Greenshank[g] | Nightjar[r] |
| Pink-footed Goose | Bittern[r] | Wood Sandpiper | Swift |
| Greylag Goose | Spoonbill | Redshank | Kingfisher |
| Barnacle Goose | Black-necked Grebe | Snipe | Kestrel |
| Brent Goose | Honey Buzzard | Great Skua | Shorelark |
| Shelduck | Marsh Harrier | Black Guillemot | House Martin |
| Wigeon | Montagu's Harrier | Razorbill | Willow Warbler |
| Gadwall | Osprey | Guillemot | Dartford Warbler |
| Teal | Spotted Crake | Little Tern | Short-toed Treecreeper |
| Mallard | Crane | Sandwich Tern | Dipper[g] |
| Pintail | Stone-curlew | Common Tern | Common Redstart |
| Garganey | Avocet | Arctic Tern | Dunnock |
| Shoveler | Oystercatcher | Black-headed Gull | Meadow Pipit |
| Eider | Grey Plover | Mediterranean Gull | Water Pipit |
| Goldeneye | Bar-tailed Godwit | Common Gull | Bullfinch |
| Smew | Turnstone | Lesser Black-backed Gull | Mealy Redpoll[g] |
| Quail | Knot | Yellow-legged Gull | Scottish Crossbill |
| Red Grouse | Curlew Sandpiper[g] | Caspian Gull[na] | Parrot Crossbill |
| Black-throated Diver | Sanderling[g] | Iceland Gull | Snow Bunting |
| Great northern Diver | Dunlin[r] | Glaucous Gull | Lapland Bunting |
| Fulmar | Purple Sandpiper | Great Black-backed Gull | Reed Bunting |

r - species on the Red list previously, g - species on the Green list previously, na - not assessed previously

# About the author

Charlie Elder is a journalist and author of the wildlife books Few And Far Between, which describes his adventures in search of Britain's rarest and most endangered animals, and While Flocks Last, an account of his travels in search of the UK's most threatened birds. He has written articles on British nature for national newspapers and magazines.

# Further reading and websites


Barnes, Simon. 2005. *A Bad Birdwatcher's Companion*. Short Books, London.

Barnes, Simon. 2012. *Birdwatching With Your Eyes Closed*. Short Books, London.

Birkhead, Tim. 2012. *Bird Sense*. Bloomsbury Publishing, London.

Brown, Andy and Grice, Phil. 2005. *Birds in England*. T & A D Poyser, London.

Cocker, Mark and Mabey, Richard. 2005. *Birds Britannica*. Chatto and Windus, London.

Couzens, Dominic. 2011. *Garden Bird Confidential*. Hamlyn, Octopus Publishing, London.

Couzens, Dominic. 2006. *Secret Lives of British Birds*. Bloomsbury Publishing, London.

Greenoak, Francesca. 1997. *British Birds, their Folklore, Names and Literature*, Christopher Helm, Bloomsbury Publishing, London.

Hayman, Peter and Burton, Philip. 1976. *The Birdlife of Britain*, Mitchell Beazley, Octopus Publishing, London.

Holden, Peter and Cleeves, Tim. 2014. *RSPB Handbook of British Birds* 4th edition, Bloomsbury Publishing, London.

Hume, Rob. 2013. *Birdwatching for Beginners*. Dorling Kindersley, London.

Hume, Rob. 2015. *RSPB Complete Birds of Britain and Europe*. Dorling Kindersley, London.

Moss, Stephen. 2016. *Do Birds Have Knees?* Bloomsbury Publishing, London.

Moss, Stephen. 2008. *How to Birdwatch*. New Holland, London.

Mullarney, Killian, Svensson, Lars, Zetterström, Dan and Grant, Peter. 2010. *Collins Bird Guide*. HarperCollins, London.

Oddie, Bill. 2013. *Bill Oddie's Introduction to Birdwatching*, Bloomsbury Publishing, London.

Sterry, Paul and Stancliffe, Paul. 2015. *Collins BTO Guide to British Birds*, HarperCollins, London.

Taylor, Marianne. 2012. *RSPB British Birdfinder*. Bloomsbury Publishing, London.

Taylor, Marianne. 2010. *RSPB British Birds of Prey*. Bloomsbury Publishing, London.

Taylor, Marianne. 2014. *RSPB Seabirds*. Bloomsbury Publishing, London.

Tudge, Colin. 2009. *The Secret Life of Birds*. Penguin, London.

*Book of British Birds*, Reader's Digest

*Bird Watching* magazine

*Birdwatch* magazine

RSPB *Nature's Home*


## Useful websites

RSPB website
www.rspb.org.uk

BTO website
www.bto.org

Wildfowl and Wetlands Trust website
www.wwt.org.uk

Wildlife Trusts website
www.wildlifetrusts.org

# Index

# Index

# Photograph credits

Bloomsbury Publishing would like to thank the following for providing photographs and for permission to reproduce copyright material within this book. While every effort has been made to trace and acknowledge all copyright holders, we would like to apologise for any errors or omissions, and invite readers to inform us so that corrections can be made to future editions.

**Key to page positions** t = top; l = left; r = right; b = bottom; ti = top inset; li = left inset; ri = right inset; tl = top left; tcl = top centre left; tc = top centre; tcr = top centre right; tr = top right; cl = centre left; c = centre; cr = centre right; bl = bottom left; bc = bottom centre; br = bottom right.

**Abbreviated photo agency names**:
BP = Birdphoto; Getty = Getty Images; RSPB = RSPB Images; SS = Shutterstock.

**Front cover:** ISBN 9781472941176 Paul Sawer/RSPB; ISBN 9781472957689 Mark Hamblin/Getty; **back cover:** tcl Chris Gomersall/RSPB; tc Mike Lane/RSPB; tr Oliver Smart/RSPB; bl Andrew Parkinson/RSPB; cl Peter Cairns/RSPB; br Genevieve Leaper/RSPB.
**Pages: 1** tl Mark Bridger/SS; tc Tobyphotos/SS; tr Marcin Perkowski/SS; cl Menno Schaefer/SS; cr Paul Miguel/FLPA; bl Gabriela Insuratelu/SS; bc ArCaLu/SS; br Giedriius/SS; **3** Drakuliren/SS; **4** l Alan Tunnicliffe Photography/Getty; tc MikeMcKen; tr Steve Dyke/EyeEm/Getty; br Daniele Carotenuto Photography; **5** tl Milan Zygmunt/SS; c Rod Teasdale; tr Nature Bird Photography/SS; bl Adam Sharp Photography/SS; br Piotr Krzeslak/SS; **6** Mark Beton/England/Alamy; **8** t Robert Bannister/Alamy; c Erni/SS; b Studioimagen73/SS; **9** t Paul Sawer/FLPA; c Duncan Usher/Minden Pictures/FLPA; b Ralph Martin/Minden Pictures/FLPA; **10** t Marek R. Swadzba/SS; c Richard Tadman/Alamy; b Brian Pollard/Alamy; **11** t Bearacreative/SS; c Kletr/SS; b Marek Cech/SS; **12** tl Jan Wegener/Minden Pictures/FLPA; li Nigel Dowsett/SS; ri Wildlife World/SS; c 2xWilfinger/SS; b Piotr Krzeslak/SS; **13** tl Vishnevskiy Vasily/SS; tr Paul Sawer/FLPA; b Em-Jott/SS; **14** all Nature Bird Photography/SS; **15** t Edo Schmidt/Alamy; c Abi Warner/SS; b Wildlife World/SS; **16** t chris2766/SS; c Erni/SS; b David Tipling/BP; **17** tl David Tipling/BP; c David Hosking/FLPA; bl Abi Warner/SS; br Tony Mills/SS; **18** t Greg Oakley/Minden Pictures/FLPA; ti Mike Lane/RSPB; c Paul Vinten/SS; b Stephan Rech/Imagebroker/FLPA; **19** c Major Wildlife/Alamy; b Soru Epotok/SS; **20** t Peter Jeavons/Alamy; c Richard P Long/SS; b Joe Gough/SS; **21** tl Do Van Dijck/Minden Pictures/FLPA; tr Neil Bowman/FLPA; b Wildlife World/SS; **22** t Steven Hogg/SS; c Dick Hoogenboom/Nature in Stock/FLPA; b Alius Imago/SS; **23** t Maria Gaellman/SS; c Rudmer Zwerver/SS; b Mark Caunt/SS; **24** t Rudmer Zwerver/SS; ti Martin Fowler/SS; c TTphoto/SS; bl Oliver Richter/Minden Pictures/FLPA; **25** tl Rebecca Cole/Alamy; tr Mark Medcalf/SS; b Anton Mizik/SS; **26** t Edo Schmidt/Alamy; c Jon_Clark/SS; bl Martin Mecnarowski/SS; **27** tl Vladimir Chernyanskiy/SS; tr geertweggen/SS; br Robert L Kothenbeutel/SS; **28** t Tobyphotos/SS; ti Mike Lane/FLPA; c Tim Gainey/Alamy; bl Arterra/Contributor/Getty; **29** tl Grant Glendinning/Alamy; tr Our Wild Life Photography/Alamy; br Mike Lane/FLPA; **30** t Bildagentur Zoonar GmbH/SS; ti bszef/SS; c Chris McLoughlin Wildlife and Nature Photography/Alamy; bl Robert Canis/FLPA; **31** t Gary K Smith/FLPA; b Martin Pateman/SS; **32** t Jesus Giraldo Gutierrez/SS; ti Marcin Perkowski/SS; c Martin Mecnarowski/SS; b 32 John Hawkins/FLPA; **33** tl Robin Chittenden/Alamy; c Derek Middleton/FLPA; br Erni/SS; **34** t John Navajo/SS; ti Wildlife World/SS; c Erni/SS; bl Toby Houlton/Alamy; **35** t Erni/SS; c Phillip Cull/Alamy; b David Tipling/BP; **36** t AlekseyKarpenko/SS; ti John Hawkins/FLPA; c Neil Walker/SS; b Marcin Perkowski/SS; **37** t Menno Schaefer/SS; c Hubert Schwarz/SS; b Erni/SS; **38** t Mike Lane/Alamy; ti Paul Sawer/FLPA; c Allanw/SS; b Abi Warner/SS; **39** t David Tipling/BP; c Paul Sawer/FLPA; b Tony Hamblin/FLPA; **40** t Alexander Erdbeer/SS; c Bearacreative/SS; b Neil Bowman/FLPA; **41** tl David Tipling/BP; tr Erni/SS; b Karel Bartik/SS; **42** all David Tipling/BP; **43** t Guentermanaus/SS; c Paul Sawer/FLPA; b Erni/SS; **44** t David Tipling/BP; ti Ondrej Prosicky/SS; c Zoltan Major/SS; bl Xpixel/SS; **45** t Mike Jones/FLPA; c David Tipling/BP; bl Allanw/SS; br Colin Robert Varndell/SS; **46** t Dave Pressland/FLPA; c David Tipling/BP; **47** t Zacarias Pereira da Mata/SS; c Erni/SS; b Rudmer Zwerver/SS; **48** t AlekseyKarpenko/SS; c Dennis Jacobsen/SS; b Rudmer Zwerver/SS; **49** t Taviphoto/SS; c Abi Warner/SS; b John Navajo/SS; **50** t Fabrizio Moglia/Getty; ti Tony Hamblin/FLPA; c Gyro Photography/Getty; b Loop Images/Contributor/Getty; **51** t Erni/SS; b Fabrizio Moglia/Getty; **52** t Bildagentur Zoonar GmbH/SS; ti Bildagentur Zoonar GmbH/SS; c Hugh Clark/FLPA; b Derek Middleton/FLPA; **53** t Richard Steel/BIA/Minden Pictures/FLPA; 53 c MiQ/SS; b Martin Mecnarowski/SS; **54** t Mark Medcalf/SS; m Marcin Perkowski/SS; b Peter Entwistle/FLPA; **55** t Erica Olsen/FLPA; Ysbrand Cosjin/SS; b Paul Hobson/FLPA; **56** t Michal Pesata/SS; c Maciej Olszewski/SS; b Richard Revels/RSPB; **57** t Maciej Olszewski/SS; c Paul Sawer/FLPA; b Abi Warner/SS; **58** t Erni/SS; c John Navajo/SS; b David Tipling/FLPA; **59** t Marcin Perkowski/SS; b Gallinago_media/SS; **60** t John Navajo/SS; c Pascale Gueret/SS; b Gallinago_media/SS; **61** t taviphoto/SS; c S Fraser/SS; b Erni/SS; **62** t Allanw/SS; ti Richard Steel/BIA/Minden Pictures/FLPA; c John Hawkins/FLPA; b Fireglo/SS; **63** t Simon Litten/FLPA; c Zakharov Aleksey/SS; b Piotr Krzeslak/SS; **64** Emjay Smith/SS; **66** t Erni/SS; ti Neil Bowman/FLPA; c Drakuliren/SS; b Garmoncheng/SS; **67** t Erni/SS; c Erica Olsen/FLPA; b Allanw/SS; **68** t John Navajo/SS; ti Martin Pelanek/SS; c Abi Warner/SS; b Bill Coster/FLPA; **69** t Roger Tidman/FLPA; b Targn Pleiades/SS; **70** t Bildagentur Zoonar GmbH/SS; ti Wildlife World/SS; c Erni/SS; b Adri Hoogendijk/Minden Pictures/FLPA; **71** t Wildlife World/SS; c SoopySue/Getty; b Nature Bird Photography/SS; **72** t Dennis Molenaar/SS; ti Antonio Guillem/SS; c Paul Sawer/FLPA; b Linda Lyon/Getty; **73** t Peter Turner Photography/SS; c Steve Round/RSPB; br Teunis Bakker/SS; **74** t Davy Veelaert/SS; ti Gertjan Hooijer/SS; c Andrew M. Allport/SS; bl Nigel Blake/RSPB; **75** tl IanRedding/SS; c Gerdzhikov/SS; br KOO/SS; **76** t Wildlife World/SS; c Abi Warner/SS; b Wildlife World/SS; **77** tl Menno Schaefer/SS; tr Duncan Shaw/Getty; b Erni/SS; **78** t Erni/SS; ti Galabin Vasilev Asenov/SS; c Erni/SS; b Erni/SS; **79** t andrekoehn/SS; c Hugh Clark/FLPA; b Emilio100/SS; **80** t Mark Medcalf/SS; c Wildlife World/SS; b KOO/SS; **81** tl Sergei Brik/SS; c Cat Downie/SS; b Vitaly Ilyasov/SS; **82** t Piotr Krzeslak/SS; ti Wildlife World/SS; c Florian Andronache/SS; b KOO/SS; **83** tl Andrew M. Allport/SS; tr Francesco de Marco/SS; b Jesus Giraldo Gutierrez/SS; **84** t Marcin Perkowski/SS; ti Erni/SS; c Colin Robert Varndell/SS; b Andrew M. Allport/SS; **85** t Borislav Borisov/SS; b Graham Braid/SS; **86** t Erni/SS; ti Romuald Cisakowski/SS; c Gerdzhikov/SS; b Drakuliren/SS; **87** cl Drakuliren/SS; tr Nick Vorobey/SS; br Paolo-manzi/SS; **88** Des Ong/FLPA; **90** t David Tipling/BP; ti Oxford Scientific/Getty; c David Tipling/BP; b Blickwinkel/Alamy; **91** t Paul Nash/SS; c Anastasia Rutkovskaya/SS; b Nick Vorobey/SS; **92** t Erni/SS; ti Gallinago_media/SS; c Abi

# Photograph credits

Warner/SS; b David Tipling/RSPB; **93** t Vishnevskiu Vasily/SS; c Nature Picture Library/Alamy; b Sander Meertins Photography/SS; **94** t D Zingel Eichhorn/FLPA; c Gallinago_media /SS; b Gallinago_media/SS; **95** t Roger Tidman/FLPA; b Abi Warner/SS; **96** Phil McLean/FLPA; **98** tl Karin Jaehne/SS; tr Karel Bartik/SS; bl Erni/SS; **99** t Dennis Jacobsen/SS; b Chris Hill/SS; **100** t Menno Schaefer/SS; ti Cowboy54/SS; c Lisa Geoghegan/Alamy; b Andrew M. Allport/SS; **101** t Davemhuntphotography/SS; c Michel Geven/Buiten-beeld/Minden Pictures/FLPA; b Paul Hobson/FLPA; **102** t Chris Hill/SS; c Sandra Standbridge/SS; b Red ivory/SS; **103** t Sue Robinson/SS; b Shaftinaction/SS; **104** t Marcin Perkowski/SS; c Piotr Krzeslak/SS; Rob Christiaans/SS; **105** tl Andrew M. Allport/SS; c Ondrej Prosicky/SS; b Pascal Halder/SS; **106** t Ondrej Prosicky/SS; c Ben Queenborough/SS; b David Tipling/Getty; **107** t Richard Whitcombe/SS; c Mark Bridger/SS; b Piotr Krzeslak/SS; **108** tl Alan Tunnicliffe Photography/Getty; cr Hans Lang/Getty; b Mark Medcalf/SS; **109** c Jackie Bale/Getty; b Andrew M. Allport/SS; **110** Gillian Pullinger/Alamy; **112** t Martin Fowler/SS; c Aleksandr Abrosimov/SS; b Geza Farkas/SS; **113** t Lhi/SS; c Smspsy/SS; b Erni/SS; **114** t Erni/SS; c Nigel Dowsett/SS; b Erni/SS; **115** c Brian Guest/SS; b Erni/SS; **116** t Martin Fowler/SS; c Bildagentur Zoonar GmbH/SS; b Martin Prochazkacz/SS; **117** tl Maciej Olszewski/SS; tr Pascal Halder/SS; b Dennis Jacobsen/SS; **118** t Paul Sawer/FLPA; c Alan Williams/Alamy; b Dennis Jacobsen/SS; **119** tr Nick Upton/RSPB; cl Guy Rogers/RSPB; b Paul Sawer/FLPA; **120** t ArCaLu/SS; ti Luca Nichetti/SS; c Dennis Jacobsen/SS; b Mark Medcalf/SS; **121** c Maksimilian/SS; b Ainars Aunins/SS; **122** t Miroslav Hlavko/SS; ti Michal Hykel/SS; c FotoRequest/SS; **123** tl Sandi Cullifer/SS; cr Targn Pleiades/SS; b V. Belov/SS; **124** t Andrew Sproule/Getty; ti Wildlife World/SS; c Ger Bosma Photos/SS; **125** tl Roland Ijdema/SS; cr Menno Schaefer/SS; b Simonas Minkevicius/SS; **126** t Maciej Olszewski/SS; ti Simonas Minkevicius/SS; c Maciej Olszewski/SS; b Erni/SS; **127** tl Ian D Nicol/SS; c MAC1/SS; b Natural Imaging/SS; **128** Papilio/Alamy; **130** t Rudmer Zwerver/SS; ti Aaltair/SS; c Paul Sawer/FLPA; b John Watkins/FLPA; **131** t Matej Ziak/SS; c Ger Bosma Photos/SS; b Matteo Photos/SS; **132** t Mark Medcalf/SS; c Abi Warner/SS; b Abi Warner/SS; **133** tl Gerdzhikov/SS; tr Terry Whittaker/FLPA; b Smishonja/SS; **134** t KOO/SS; ti Erni/SS; c IanRedding/SS; b Mark Medcalf/SS; **135** t YK/SS; b Nigel Blake/RSPB; **136** t Menno Schaefer/SS; ti Rock Ptarmigan/SS; c Malcolm Schuyl/FLPA; b julylx/SS; **137** t Richard Constantinoff/SS; b Rock Ptarmigan/SS; **138** t Erni/SS; ti Luka Hercigonja/SS; c Erica Olsen/FLPA; **139** t Cristian Gusa/SS; b Mark Caunt/SS; **140** t Erni/SS; c Erni/SS; b Martin Fowler/SS; **141** tr Roger Tidman/FLPA; c Erni/SS; b Erni/SS; **142** tl Gabriela Insuratelu/SS; cr Karel Bartik/SS; b Mike Lane/FLPA; **143** t Pascal Halder/SS; b FotoRequest/SS; **144** Erica Olsen/FLPA; **146** t V. Belov/SS; c John Lawson/Getty; b Drakuliren/SS; **147** tl Imagebroker/FLPA; c Aaltair/SS; b Erni/SS; **148** t Bildagentur Zoonar GmbH/SS; c Timbobaggins/SS; b KOO/SS; **149** t Arcticphotoworks/SS; c Menno Schaefer/SS; b Tony Mills/SS; **150** t Chris Gomersall/Alamy; ti Arterra Picture Library/Alamy; c Abi Warner/SS; b Keith M Law/Alamy; **151** t Steven Ruiter/Nature in Stock/FLPA; c Wil Meinderts/Minden Pictures/FLPA; b Duncan Shaw/Getty; **152** t Rob Christiaans/SS; ti TheNatureWeb.Net/SS; c Astrid Kant, Buiten-beeld/Minden Pictures/FLPA; b Steve Round/RSPB; **153** t FotoRequest/SS; c P.V.R.M/SS; b KOO/SS; **154** t Mark Caunt/SS; c Richard A. Evans/SS; b Dennis Jacobsen/SS; **155** t Dave Montreuil/SS; c Mateusz Sciborski/SS; b Erni/SS; **156** ti WayneDuguay/SS; t Marcin Perkowski/SS; c Ernie Janes/SS; **157** t Elliotte Rusty Harold/SS; b Erni/SS; **158** t Erni/SS; ti Paul Reeves Photography/SS; c Abi Warner/SS; b Maciej Olszewski/SS; **159** t Anton Mizik/SS; c Abi Warner/SS; **160** t Tobyphotos/SS; c RazvanZinica/SS; b Dennis Jacobsen/SS; **161** tl Daniele Occhiato/Minden Pictures/FLPA; tr Hugh Clark/FLPA; br xpixel/SS; **162** t Marco Rolleman/SS; ti Astrid Kant/Minden Pictures/FLPA; c Paul Broadbent/SS; b Wildlife World/SS; **163** t Bohus Cicel/SS; b Martin Fowler/SS; **164** t Pedrosanmar/SS; ti Wolfgang Kruck/SS; c NatureMomentsuk/SS; b Maksimilian/SS; **165** t Erni/SS; b vagabond54/SS; **166** Fotosearch/Getty; **168** t Christian Musat/SS; ti Christian Musat/SS; c Paul Reeves Photography/SS; b Maciej Olszewski/SS; **169** t David Fowler/SS; b Dave Montreuil/SS; **170** t Nature Bird Photography/SS; ti AlekseyKarpenko/SS; c Ernie Janes/RSPB; b Paul Broadbent/SS; **171** t poganyen/SS; c Ray Kennedy/RSPB; b RazvanZinica/SS; **172** t Wolfgang Kruck/SS; ti Roland Ijdema/SS; c Wolfgang Kruck/SS; b Genevieve Leaper/RSPB; **173** t Mateusz Sciborski/SS; b Leif Ingvarson/SS; **174** t Erni/SS; c Ondrej Prosicky/SS; b Olaf Kruger/Getty; **175** t Targn Pleiades/SS; b Ed Marshall/RSPB; **176** t Gert-Jan IJzerman/NIS/Minden Pictures/FLPA; ti Wilfried Martin/Getty; c Fotosearch/Getty; b Fotosearch/Getty; **177** t Menno Schaefer/SS; b Guy Rogers/RSPB; **178** t AndreAnita/SS; ti Gertjan Hooijer/SS; c Paul Reeves Photography/SS; **179** t Andreas Rose/SS; b Maria Gaellman/SS; **180** t Erni/SS; c A. S. Floro/SS; b Abi Warner/SS; **181** t Erni/SS; b Michal Pesata/SS; **182** t Stephan Rech/Getty; c Jerome Whittingham/SS; **183** tl Frans Sellies/Getty; tr Stephan Rech/Getty; b Rudmer Zwerver/SS; **184** t Luca Nichetti/SS; ti NatureMomentsuk/SS; c Giedriius/SS; b Southern Lightscapes-Australia/Getty; **185** tl Eric Isselee/SS; tr Mark Medcalf/SS; b Erni/SS; **186** Ian_Sherriffs/SS; **188** Penny Cross; **190** Franke de Jong/SS.


# Acknowledgements

With many thanks to Julie Bailey, Senior Commissioning Editor at Bloomsbury, Susan Smith, MBA Literary Agents, the RSPB, designer Rod Teasdale, editor Louise Morris and to family and friends for their support and encouragement during the writing of this book.